ISLAND IN THE NET

ISLAND IN THE NET

Digital Culture in Post-Castro Cuba

Steffen Köhn

Illustrated by Nestor Siré

Princeton University Press

Princeton and Oxford

Published by Princeton University Press
41 William Street, Princeton, New Jersey 08540
99 Banbury Road, Oxford OX2 6JX

press.princeton.edu

GPSR Authorized Representative: Easy Access System Europe—
Mustamäe tee 50, 10621 Tallinn, Estonia, gpsr.requests@easproject.com

All Rights Reserved

ISBN 9780691273136
ISBN (pbk.) 9780691273143
ISBN (e-book) 9780691273181

Library of Congress Control Number: 2025939527

British Library Cataloging-in-Publication Data is available

Editorial: Fred Appel and James Collier
Production Editorial: Jaden Young
Text and Cover Design: Heather Hansen
Production: Lauren Reese
Publicity: William Pagdatoon
Copyeditor: Joseph Dahm

Jacket images: Photographs by Nestor Siré

This book has been composed in Adobe Text Pro
with Varietta and Calling Code

1 3 5 7 9 10 8 6 4 2

CONTENTS

ACKNOWLEDGMENTS

I am writing these lines from Havana, where I am currently engaged in a new research project, just two weeks into Donald Trump's second term. The economic and infrastructural challenges facing Cuba today are even more severe than during the period of my initial fieldwork for this book. In October, the national power grid collapsed multiple times, leaving the entire island without electricity for days. Food prices have risen significantly, shortages of essential goods, fuel, and medicine are dramatic, and the government has just announced a "partial dollarization" in its latest attempt to steer a completely bankrupt economy. The outlook remains grim, particularly under the new U.S. administration, with Marco Rubio, an aggressive Cuba hardliner, serving as secretary of state. His policies are likely to intensify the already harsh measures against Cuba, while the new administration's antimigration stance will undoubtedly make the migratory endeavors of the many Cubans who chose to leave the island more difficult and perilous. Amid these challenges, I am continually struck by the resilience, creativity, and grace with which my Cuban friends and acquaintances navigate their lives and pursue their projects. Their perseverance is a profound source of inspiration.

This book would not have been possible without the generous contributions of numerous individuals and institutions. I am deeply indebted to all those who offered their time, resources, and intellectual support throughout the research and writing process.

First and foremost, I extend my heartfelt gratitude to Nestor Siré, whose friendship, collaboration, and creative insights have been integral to this project. I am also profoundly thankful to the Instituto Cubano de Investigación Cultural Juan Marinello, which provided an intellectual home for me in Havana and facilitated this research. Special thanks go to Maybel Martínez Rodríguez, Marcia Peñalver, Hamlet López, and the late Rodrigo Espina Prieto, whose guidance and advice were indispensable.

I am immensely grateful to the colleagues who invited me to present the book's arguments at their institutions during various stages of this project as well as for the responses and insights of the audiences. My sincere thanks to Eva

Otto at the University of Copenhagen, Suzana Ramos Coutinho at the Universidade Presbiteriana Mackenzie in São Paulo, the Multimedia Anthropology Lab at UCL in London, Tobias Días at Kunsthal Aarhus, Frank Heidemann and Sahana Udupa at LMU Munich, Sam Hopkins at KHM Cologne, Markus Beckedahl, Carolin Moje, and Paolina Wandruszka at re:publica Berlin, Matthias Krings at Johannes Gutenberg University Mainz, and Ariel Camejo at Universidad de La Habana.

I also wish to acknowledge the art institutions that exhibited and supported the research-based artworks that accompany this book. My thanks to Bartek Frąckowiak, Anna Galas-Kosil, Ewa Kozik, and Paweł Wodziński at Biennale Warszawa; Andreas Zingerle at Mur.at; the team at the Taiwan International Video Art Exhibition, including Zoe Chia-Jung Yeh, Shih-yu Hsu, and Jessica Chien; Jon Uriarte, Sam Mercer, Arieh Frosh, and Ioanna Zouli at the Photographers' Gallery in London; Johan Ahrenfeldt, Torbjørn Eriksen, Ole Nielsen, and Anders Emil Rasmussen at Moesgaard Museum Aarhus; Maja Burja, Luka Frelih, and Anže Zorman at Ljudmila Art and Science Laboratory Ljubljana; and Marcela Okretič and Janez Fakin Janša at Aksioma Institute for Contemporary Art in Ljubljana. Further, I am deeply grateful to Ines Neidhardt at the German Embassy in Havana as well as Michael Thoss and Lena Brode at the Goethe Institute Havana for their invaluable assistance. I also thank the Knud Højgaard Foundation, the Aage and Johanne Louis-Hansen Foundation, the Augustinus Foundation, the Beckett Foundation, and Aarhus University for their generous funding. My research was generously sponsored by the Deutsche Forschungsgemeinschaft (DFG, German Research Foundation) under project number 428086777. Additionally, the publication of this book was supported through a kind grant from the Aarhus University Research Foundation.

I owe a special debt of gratitude to Fred Appel of Princeton University Press for his unwavering support of this book from its inception as well as to series editors Bill Maurer and Tom Boellstorff for their early interest in the project. Brenda Chalfin and Heidi Härkönen provided generous reviews as well as invaluable feedback and suggestions, while Joseph Dahm's editorial expertise greatly enhanced the book. Any errors are of course my own.

I am deeply thankful to the friends and colleagues whose assistance, ideas, and insightful exchanges have significantly enriched this work. My thanks to Hope Bastian, Humberto Calzada, Jorge Duany, Undine Frömming, Faye Ginsburg, Guillermo Grenier, Katrin Hansing, Bert Hoffmann, LZ Humphreys, Alejandro Jaramillo, Eliecer Jiménez, Antoni Kapcia, Andy Lawrence, Nicolas Malevé, Henrike Naumann, Petra Nováčková, Anna Cristina Pertierra, Juan C. Toledano Redondo, Ariana Reguant, Gaia Tedone, Clemens Villinger, and Julia Weist.

At the Department of Anthropology at Aarhus University, I am grateful to my colleagues for fostering such a supportive and stimulating environment for making, teaching, and debating anthropology. Special thanks to Kenni Hede, Susanne Højlund, Ton Otto, Andreas Roeppsdorf, Carolina Sánchez Boe, Christian Suhr, Heather Swanson, Marie Louise Tørring, Noa Vaisman, Christian Vium, Nina Holm Vohnsen, and Cameron Warner. At the Department of Digital Design and Information Studies, I thank Jussi Parikka, Kasper Shiølin, Madga Tyżlik-Carver, and Christian Ulrik for their intellectual companionship.

In Cuba, my work has been profoundly shaped by the friendship and support of Lázaro Bottino Alfonso, Yonlay Cabrera, Miguel Coyula, Lynn Cruz, Paolo De, Omar Estrada, Dina Fernández, Davicito Ferrán, David Ferrán, Johan Ferrán, Adonis Ferro, Ernesto Gamboa, Yasser González, Daniela Gutiérrez, Yoan M. Gutiérrez, Maurice Haedo, Chuli Herrera, Francesco Innocenti, Andreas Knobloch, Adriano López, Erick Maza, Amarilis Mederos, Evelio Mederos, Ever Mederos, Pavel Méndez, Erick J. Mota, Joaquín Perdomo, Mari Pérez, Eduardo Pujol, Lynet Rivero, Fidel Alejandro Rodríguez, Yainet Rodríguez, José Rubio Acosta, Katia Sánchez, Enedys Seijo Planes, Charlenys Vidal, Yoss, and many others.

I am also grateful to my friends in Germany—Alexander Bley, Johannes Büttner, Paola Calvo, Raki Fernández, Felix Girke, Patrick Jasim, Henrika Kull, Nicole Medvecka, Ginés Olivares, Helge Peters, and Andrea Sunder-Plassmann—for their camaraderie and to my parents, Karin and Wilfried Köhn, for their infinite support.

Finally, I owe my deepest gratitude to my partner, Martha Otwinowski, for standing by my side throughout this journey and for the joy she brings to my life.

Parts of chapters 3 and 4 have previously appeared as "Fragile Connections: Community Computer Networks, Human Infrastructures, and the Consequences of Their Breakdown in Havana, Cuba" in *American Anthropologist* 124, no. 2 (2022): 383–98, and "Swap It on WhatsApp: The Moral Economy of Informal Online Exchange Networks in Contemporary Cuba" in *Journal of Latin American and Caribbean Anthropology* 27, no. 1 (2022): 80–100. I am grateful to the publishers for permitting me to expand on these materials.

This book is dedicated to the creative communities I have had the privilege to collaborate with, particularly the SNET gamer community, the members of the Copincha maker space, and the compilers, distributors, and users of *el paquete semanal*. Their ingenuity and persistence continue to inspire me.

A GUIDE TO THE MULTIMODAL COMPLEMENTS
OF THIS BOOK

Each of the six main analytical chapters in this book is accompanied by a work of multimodal anthropology, providing readers with vivid, immersive insights into the complexities of Cuban media realities and the diverse technological communities examined in this study. These works highlight the creative strategies employed by Cubans to compensate for limited Internet connectivity as well as the vernacular infrastructures they have developed to navigate these challenges. They span a variety of representational formats, including a video game, two interactive installations, a live video art piece, an ethnographic documentary, and an expanded cinema installation. Descriptions of these anthropological media and their creation are presented as interludes following each chapter, and readers can access the works directly via QR codes. Since not all these projects can be adequately represented on a single screen, these QR codes do not link to a streaming platform like Vimeo or YouTube. Instead, they direct users to a data package stored in a cloud space, which contains the work itself (whenever feasible) and/or extensive documentation and supplementary materials, such as installation views, video documentation, and software code. The structure of these data packages—organized into folders and subfolders—deliberately mirrors the emic Cuban approach to data organization found in grassroots media distribution networks like *el paquete semanal* (the weekly package) or SNET.

Most of these works are the result of my long-standing collaboration with Cuban media artist Nestor Siré. Embracing an anthropology of cocreation, we developed these projects in partnership with local software developers, science fiction writers, social media influencers, and makers of alternative infrastructures. As such, they emerge from—and invite audiences into—the intricate media worlds inhabited by my research participants. These works have been exhibited at art events and venues worldwide, including the Warsaw Biennale, Aksioma Institute for Contemporary Art in Ljubljana, the Photographers' Gallery in London, Moesgaard Museum in Aarhus, Hong-Gah Museum in

Taipei, and Havana Espacios Creativos. With the publication of this book, they are being made accessible online for the first time.

Nestor has also contributed to this book by crafting its distinctive visual style. For each chapter and interlude, he has created collages composed of images that document the social and material practices described in the text. While these appear in black-and-white within the book, they can be viewed in full color via these QR codes, further enriching the reader's experience.

ISLAND IN THE NET

Cuba's Digital Awakening

On July 11, 2021, Cuba erupted in its largest outbreak of public discontent in decades. Thousands of people took to the streets to voice a variety of grievances that had been degrading the quality of life on the island for years, such as increasingly severe power outages and dramatic shortages of food, fuel, and medicines. Growing dissatisfaction had been simmering under the surface for some time, with groups of young artists and musicians organizing several small protests against increasing censorship, until the worsening coronavirus pandemic and its apparent mishandling by the Cuban government culminated in a sudden outburst of public outrage. Demonstrations are a rare occurrence in Cuba's tightly controlled society, and the speed with which they spread, washing over all major cities within a couple of hours, was an entirely new phenomenon. It was the most visible expression yet of a social transformation that is irrevocably changing state-society relationships on the island: Cuba's digital awakening.

After endless hesitation, the Cuban state had finally begun enabling citizens' access to mobile Internet a mere two and a half years before the incidents of July 11. In a country where the government still controls all mass media, the protests of that day would not have become a nationwide event without the mass proliferation of social media. Videos of the riots, which began in the small town of San Antonio de los Baños, about an hour outside Havana, quickly spread through Facebook and on WhatsApp and Telegram groups, reaching large segments of the population and spurring similar protests across the country until the government shut down the Internet completely in the early afternoon.

July 11 revealed both the transformative power and the limits of networked digital technologies in an authoritarian system. While social media has opened up new opportunities for Cuban citizens to organize in unprecedented ways outside of state-orchestrated organizations and state-controlled media, the Cuban government quickly regained control over the streets because the protests weren't backed by an organized opposition. In the weeks after the incident, the government addressed some of the protesters' bread-and-butter concerns but also put hundreds of demonstrators in jail. This government crackdown, combined with continued political standstill and the unresolved economic crisis, led to an unprecedented mass exodus, particularly of young and aspiring Cubans, in the months and years that followed. This was in turn facilitated by social media, as people shared information online about migration routes and immigration regulations.

While Cuba's digital revolution might not have brought about systemic political change, this book explores how Cubans' increasing access to digital technologies and the Internet are fundamentally reconfiguring the power dynamics between citizenry and government by allowing Cuban people to become more autonomous from state structures. The Cuban government had long sought to evade the erosion of its media monopoly by restricting its citizens' access to digital technology and the Internet due to fears of freedom of information and expression. As a result, the island was one of the least connected countries in the world until just a few years ago. Since the institutionalization of the state socialist model in 1961, two years after the triumph of the Revolution in 1959, it has maintained a tight grip on the public sphere, closing all independent spaces where people could potentially publicly express dissent. Cuban authorities expanded access to the Internet and digital communication technologies only when it became clear that abandoning them would be ruinous for the economic survival of the island's economy, which is structurally heavily dependent on exports and tourism.

Throughout his lifetime, Fidel Castro himself remained ambivalent about the Internet, sometimes praising its potential as a powerful tool in the hand of the revolutionary, while at the same time warning that unrestricted access would inevitably change Cuban society (Hoffmann 2004; Recio Silva 2013). The Cuban government was therefore cautious and only gradually made the Internet more available while desperately trying to defend its revolutionary project. Until 2013, Internet cafés in the better tourist hotels were (at least officially) the only places where users could connect via satellite links at exorbitant hourly rates. But that year, a fiber-optic cable to Venezuela finally integrated the island into the global broadband architecture. Starting in summer 2015, the government began installing Wi-Fi hotspots in public places, first in Havana and then in other cities. It was through these public hotspots, which were still just barely affordable, that the vast majority of Cubans first encountered the Internet. Three and a half years later, in December 2018, the state rolled out a 3G cellular network that eventually provided mobile Internet for smartphone users. In July 2019, it was updated to the LTE standard. The opening of the

Wi-Fi parks had changed the way Cubans on the island could communicate with the world and particularly with their relatives in the diaspora (initially via Facebook or the audiovisual telephony and instant messaging software imo). The introduction of mobile data now changed the way Cubans communicated with each other, as messaging groups on WhatsApp or Telegram became important sites of socialization.

Although the Cuban Internet is still characterized by slow speeds, high costs, and inadequate infrastructure, Cubans have continued to find inventive ways to take advantage of the increasingly available technology and develop strategies to compensate for the lack of Internet connectivity and affordability. For many years, technology enthusiasts have created vernacular infrastructures that make up for the limitations of the state-supported digital infrastructure by extending, bypassing, or replacing it, such as grassroots community computer networks or "sneakernets" in which digital data are transferred by physically carrying portable hard drives and memory sticks. Mutually reinforcing each other, the state's tentative liberalization policy and its citizens' inventiveness have therefore come to endanger the Communist Party's control over the economy, politics, social life, and media usage. Cuba's former authoritarian public sphere (Dukalskis 2017) has transformed into an emerging networked public sphere encompassing various autonomous and semiautonomous social and communicative spaces that offer Cubans new political and economic opportunities.

Island in the Net examines these ongoing changes, detailing how citizens are harnessing digital technology to create new alternative support infrastructures and thriving digital black markets for trading scarce consumer goods and foreign currency as well as new spaces for public debate. The book's ethnographic narrative spans from the opening of public Wi-Fi hotspots in 2015 to the social-media-fueled protests of July 11, 2021, and their aftermath. It is grounded in fourteen months of fieldwork, conducted over the course of five years and spanning multiple locales. I spent ten months researching in Havana (half of that time during the height of the pandemic), the central location for both government-sponsored innovation and bottom-up parallel practices. To understand how the island's digital transformation plays out outside the capital and to chart the local vernacular technological infrastructures, I spent another total of two months in the provincial capitals of Camagüey, Ciego de Ávila, and Matanzas. To capture the transnational dimensions of Cuba's emerging networked public sphere and to follow up with the many research participants who decided to emigrate during my research, I worked for an additional month each in Miami-Dade County, the heart of the Cuban exile community, and in Madrid, a destination that became increasingly popular among a younger generation of émigrés seeking to escape the ideological polarization of the Cuban American community in the United States.

I joined my collaborators in the comfort of their homes, closely observing their consumption of both local and global media content. I observed some of them recording YouTube videos or engaging in multiplayer games on DIY computer networks. Alongside others, I participated in self-organized workshops on 3D printing, attended Otaku festivals that celebrated Japanese pop culture, and visited shops catering to remittance senders within the Cuban diaspora in Hialeah. Furthermore, I dutifully fulfilled requests to deliver technological devices, foreign cash payments I had received in someone's name on my bank account, and, increasingly, basic medications to individuals who sought my assistance.

In total, I conducted approximately a hundred formal and informal interviews with various key figures in Cuba's emergent digital culture, including novice netizens, connection brokers at public Wi-Fi parks, admins and users of vernacular technological infrastructures, social media personalities, content creators, software developers, makers, and many others. Whenever possible, I held these interviews in the authentic settings where the participants carried out their activities. For instance, I climbed with some of them onto rooftops as they skillfully mounted new Wi-Fi antennas or followed others through their neighborhoods as they delivered hard drives brimming with data to their customers.

Given that some of my research participants are public figures in Cuba, such as social media influencers, entrepreneurs, writers, or artists, and many of them made significant contributions to the multimodal projects accompanying this publication, I have chosen not to anonymize all names by default, as I believe anonymization would be counterproductive to their interests. Carlo Cubero (2021) highlights that the ethics of anonymity are intricately linked with broader concerns related to representation, voice, and power dynamics throughout the research process. As Margot Weiss (2021) emphasizes, in many instances our interlocutors are actually cotheorists who deserve credit for their ideas and contributions to our knowledge production. Because it is paramount to me to recognize each participant's involvement in this work, I prioritized their preferences, allowing them to decide whether they wished to be identified by name or remain anonymous. In all instances, I engaged in open discussions with collaborators regarding the pros and cons of anonymization and the potential repercussions of using their real names. Generally, when referring to interlocutors by their first names, I have used pseudonyms, while their full names denote real participants. The names of businesses mentioned are real if their owners permitted their use. In cases where interlocutors privately expressed critical views of the government or were associated with the dissident movement, I took great care to remove any personally identifiable information.

⋈

The following section summarizes the broader economic and social changes that the island has experienced in recent years and that have turned the lives of Cuban citizens upside down. Cuba's digital awakening must be understood against the backdrop of developments such as the government's timid reforms of the private sector, which have legalized some formerly informal occupations while pushing others into illegality; the COVID-19 pandemic that has devastated the country's tourism-dependent economy and led to a veritable supply crisis; and a poorly timed monetary reform that resulted in rampant inflation.

The remainder of the introduction then discusses the interventions that this book makes in three interdisciplinary strands of research on the social impact of technology. First, I engage the literature on the infrastructural turn in the social sciences to provide fresh insights into how people-embedded alternative infrastructures can become a ground for contestation. I analyze how collective practices of constructing and maintaining technological systems might constitute infrapolitical acts. I unfold how such vernacular infrastructures can thus become primary sites for negotiating conflicting ideas about citizens' rights and obligations as well as state-promoted and popular values and morality.

Second, I explore how the Cuban case contributes to our understanding of the complex impacts of the Internet on citizen-state relationships in authoritarian settings. In a departure from oversimplistic grand theories about the Internet and political change that either celebrate it as a "liberation technology" (Plattner and Diamond 2012) or warn of "networked authoritarianism" (Morozov 2011) or the Internet's tendency to destroy the deliberative public sphere (Habermas 2022), I argue for a more nuanced comprehension of the possibilities and limitations of networked technologies for citizens of autocratic states. While increasing Internet access has expanded the amount of information available to ordinary people, facilitated new and important ways for citizens to discuss public issues, and provided a space for activism, long-tested "offline" forms of political propaganda and repression via the "old" mass media remain powerful and allow the state to contain online expression within well-policed boundaries.

Third, I consider how multimodal approaches to anthropology can provide a framework for recognizing the central role of media production in the everyday lives of our research participants and for creating new forms of public scholarship that can engage both nonacademic and local audiences and even change the very ways we do research. One of the main conceptual and methodological innovations of this publication is that each of its chapters is accompanied by a work of multimodal anthropology that performs the arguments of the respective chapter and was created in collaboration with some of the protagonists of Cuba's emergent digital culture. Through these "experimental collaborations" (Estalella and Sánchez Criado 2018), I have been able to transcend the traditional separation of roles between researcher and researched as well as the established approaches of distant and engaged participant observation typically associated with conventional ethnographic research.

A Country in Transition

Despite the apparent lack of overt political transformation, the past decade has been very turbulent for the island, as both internal steps toward reform as well as external events and pressures have had profound implications for Cuban politics and society. Soon after taking over power from Fidel in 2008, Raúl Castro launched a reform process to "update" the Cuban economic and social model that introduced more market mechanisms, expanded the private sector, and reduced government sector employment and state subsidies.[1] This withdrawal of the state from many of its previous functions also granted citizens greater freedom. Cubans could now lease land and buy or sell real estate, and most importantly, the government allowed the number of authorized occupations and *cuentapropistas* (nongovernmental or self-employed workers) to expand.[2] A 2012 law liberalized travel and migration and eliminated the need for a permit to enter or leave the national territory. Scholars have described these political changes of the post-Fidel era as a transformation from a "charismatic post-totalitarian regime" to a "maturing post-totalitarian" one that emphasizes economic performance as a compensatory source of legitimacy (Centeno 2017) or toward "bureaucratic socialism" (Hoffmann 2016). This transition away from charismatic politics was completed when Miguel Díaz-Canel succeeded Raúl Castro as the country's president in 2019. Díaz-Canel is the first Cuban head of state who was born after the Revolution and therefore cannot derive his authority from having participated in it.

Meanwhile, the island's relations with its archenemy, the United States, which since 1960 has maintained the most enduring and punitive trade embargo in modern history, have taken several twists and turns. The Cuban thaw during the Obama administration, which began in December 2014, saw the lifting of some U.S. travel and remittance restrictions, improved access for U.S.

banks to the Cuban financial system, and the reopening of embassies. It culminated in the first visit to the island by a U.S. president since 1928. However, that policy shift ended abruptly when Donald Trump was elected and reversed Obama's relaxed Cuba policy, reinstating and even tightening some aspects of the embargo such as restrictions on commerce with military-owned businesses (including many hotels) and on remittances and travel to Cuba by U.S. citizens.

Trump's electoral victory was one of several factors contributing to a series of economic challenges that affected the country and significantly impacted the standard of living of my research participants throughout my fieldwork. Around the same time, Cuba's options for foreign trade and investment shrunk considerably, with longtime business partner Venezuela's ongoing economic disintegration and Brazil's and Bolivia's shift to right-wing governments (two countries to which Cuba was leasing medical professionals for hard currency remuneration). The island's economy was thus already in its worst shape in two decades before COVID-19 reached Cuba in mid-March 2020. The lockdowns, border closures, and dearth of tourists in the wake of the pandemic then stifled the economy and led to a significant drop in food imports, which led to a supply crisis that affected virtually the entire population.

These external shocks were joined by an internal shock in the form of an ill-timed currency reform to unify the two national currencies, the Cuban peso (CUP) and the dollar-pegged Peso convertible (CUC). The CUC was introduced in 2004 to replace the U.S. dollar, which had become first an informal and then a second legal currency during Cuba's economic collapse in the early 1990s after the fall of its most important trading partner until then, the Soviet Union. Both currencies, the CUC and the CUP, were used in separate economic spheres. While the peso was the currency of the state socialist system, used for paying salaries and acquiring heavily subsidized basic goods through the *libreta* (ration card), the CUC served as the currency within the tourism sector and was the preferred means of payment for most private businesses. Additionally, the CUC was necessary for locals to purchase imported goods like cars, televisions, and cell phones.

As Martin Holbraad (2017) has noted, the dual currency for Cubans represented not only two different economic but also two distinct moral orders, that of socialist distribution and the capitalist market—the latter excluding a large portion of the Cuban population, namely those who do not have the privilege to receive dollar remittances from family abroad and/or who do not work in the tourist or the private sector. Moreover, it cut the link between skill level and income, as, to cite a widely used local example, a brain surgeon who would receive one of the highest CUP salaries within the state system still earned less than a waiter at a tourist hotel who received their tips in CUC. Those who had access to dollars could buy subsidized peso goods at a fraction of the market price while at the same time enjoying imported luxuries that were out of reach for large segments of society.

This breakdown of the former egalitarian reward system shattered many Cubans' belief in the revolutionary process. Since the collapse of the Soviet Union, the Cuban economy's overt reliance on tourism and remittances to generate hard currency has become the driving force behind a palpable growth in inequality that is increasingly reminiscent of prerevolutionary conditions. As Hope Bastian (2018) and Katrin Hansing and Bert Hoffmann (2020) have argued, new class divisions have emerged in recent years along clearly visible racial lines as the overwhelming majority of remittance-sending families in the United States (many of them descendants of the prerevolutionary bourgeoise) are white and send money to their white relatives. They also show that a greater proportion of Afro-Cubans are living in substandard housing in unattractive areas that are unsuitable to rent to tourists on Airbnb, another important new income generator.[3]

The reunification of both currencies became necessary not only due to the evident inequalities generated by the separate economic spheres but also to address and eliminate economic distortions.[4] While ending the dual currency system was a central promise of Raúl Castro's reform course, like many other measures it was delayed and then eventually implemented at the worst possible moment. When it took effect on January 1, 2021, this was at the height of a global economic recession, which in Cuba was exacerbated by the negative impacts of the pandemic and the tightening of sanctions by the Trump administration. An inflationary push of an estimated 500 percent in 2021 was the consequence. Even though the state adapted price and wage structures to cushion the effects on the population, living standards were seriously affected.

As the government also cut some of the subsidies for basic goods and services, electricity prices and living costs skyrocketed. Due to the severe shortage of foreign currency caused by the collapse of the tourism industry, weak exports, and low production, the fixed exchange rate of 24 CUP for 1 dollar did not reflect market conditions. As a result, the black market for foreign currencies, such as the dollar and the euro, thrived. In the first days of September 2022, twenty months after the reform, the black market exchange rate of the dollar had already climbed to the emblematic mark of 150 CUP. By May 2024, this rate had risen to 400 CUP before stabilizing at around 320 to 340 CUP in the subsequent months.

The government itself fueled the demand for dollars when at the height of the supply crisis in the summer of 2020 it opened a new chain of hard currency stores, the so-called MLC stores.[5] The idea behind these shops was to channel hard currency held by the population into the hands of the state rather than the black market. The state itself would purchase desirable high-value goods such as electronic equipment on the international market to sell to its population at lower prices than on the black market and within a legal framework. However, due to the supply shortages triggered by the pandemic, the government resorted to depleting its local currency stores to stock the shelves of the MLC stores. Hence, the MLC stores, initially established to primarily offer nonessential luxury goods, swiftly transformed into the sole locations where essential items like meat, cheese, toilet paper, and detergent could be purchased. This effectively limited access to these goods for ordinary Cubans without access to dollars or euros. Because people can make purchases at these stores only with a special bank card, this measure again widened the gap between those who had hard currency to spend, such as entrepreneurs, high-ranking government cadres, remittance recipients, successful artists, and employees of foreign companies, and those who did not.

Throughout 2021, after the monetary reform had been implemented, the peso stores became emptier and emptier and often did not sell more than bottled water, rum, and sometimes frying oil. Consequently, it became virtually impossible to purchase anything in state stores using the official national currency. People thus had to resort to exchanging their pesos received as state salaries on the black market at increasingly inflated rates in order to afford even basic necessities. Rather than achieving currency unification, this desperate government response resulted in a partial re-dollarization of the economy, where expenses were in dollars but wages remained in the national currency. As a result, Cuban families relying solely on their state income to cover expenses found themselves impoverished and struggling to make ends meet.[6] The state's lack of hard currency not only limited food imports but also led to an increased decapitalization of industries and infrastructure, the latter contributing to an escalating energy crisis. If Cubans had already grown accustomed to decaying buildings, potholed roads, and a broken transportation system, now power outages were accumulating, caused by malfunctions and failures in aging thermoelectric plants, fuel shortages, and a lack of resources for maintenance.

Compared to the economic crisis of the 1990s—euphemistically called the Período especial en tiempos de paz (Special Period in Times of Peace) by the authorities—the current ongoing economic depression is unfolding in the context of changed relations between the state and its citizens. The present government lacks the charismatic leadership of Fidel Castro, who was able to bring his full historical weight to bear when antigovernment demonstrations last erupted on August 5, 1994, at the peak of the Special Period. Fidel personally confronted protesters on Havana's Malecón, after which the up-heaval died down. A crucial difference also lies in the fact that the state's recent liberalization of Internet access and acceptance of a gradual diversification of the public sphere also means that the Cuban people nowadays have many more opportunities to self-organize via technology-supported alternative networks and more leeway to voice their discontent.

What brings Cubans together today over common concerns is not the political mass organizations such as the neighborhood Comités de Defensa de la Revolución (Committees for the Defense of the Revolution), but Facebook groups, WhatsApp chats, and Telegram channels, or collectively maintained vernacular infrastructures such as community computer networks. The 1994 demonstrations, therefore, do not serve as a template for the historic July 11, 2021, protests. This time, the protests were not limited to a single location but moved quickly throughout the country due to the mobilization power of social networks and were therefore much more difficult to contain. The government saw no other way than to cut the Internet during the demonstrations, and it reacted with a much hasher crackdown than after the so-called Maleconazo in 1994, giving lengthy prison sentences to hundreds of protesters.

The Power of Infrastructure

The July 11 protests represent the culmination of many of the developments about which this book is concerned. One of my main arguments is that the changing relationship between Cuban citizens and the state can be produc-tively understood through the lens of infrastructure. Since Michael Mann's (1984) influential conceptualization of the state's infrastructural power, schol-ars in anthropology, science and technology studies, human geography, and related disciplines have delved into the analysis of how infrastructure shapes and delineates social existence. Infrastructure is seen as a critical site for the organization and regulation of society. Researchers have examined how state-driven infrastructures have become symbols of modernity, development, and progress, while also shedding light on the dire consequences that arise when these infrastructures fail, emphasizing their significance as key locations for social and physical reproduction. They have further demonstrated how the power relations built into seemingly everyday infrastructures govern into the most intimate aspects of citizens' lives.

Investigating the social life of infrastructure in their field sites, anthropolo-gists have encountered two distinct forms of "infrastructural violence" (Rod-gers and O'Neill 2012). Researchers focusing on postcolonial urban contexts have documented the *exclusion* and violence experienced by marginalized groups in relation to access to basic infrastructure. These exclusions result from postcolonial inequalities, capitalist privatization, neoliberal neglect, and apartheid technopolitics. For instance, Filip De Boeck and Marie-Françoise Plissart (2004) have examined the effects of neoliberal policies and austerity measures on the residents of Kinshasa, revealing urban fragmentation and the breakdown of the city's social fabric. Antina von Schnitzler (2016) has studied struggles surrounding essential services like water and electricity in post-apartheid South Africa. Nikhil Anand (2017) has explored how access to water infrastructure, or the lack thereof, in postcolonial Mumbai contributes to the production of unequal forms of citizenship.

In contrast, anthropologists of socialism and postsocialist transformation have described forms of violent *inclusion* into state infrastructure as, according to Marxist materialist doctrine, the personal involvement in building public works was expected to lead to new sociomaterial relationships and, ultimately, to the creation of new socialist human subjects. In this fashion, Dimitris Dalakoglou (2012) has detailed how in socialist Albania the great majority of citizens were forced to create their socialist homeland by building many miles of new roads. Likewise, Caroline Humphrey (2005) has discussed how socialist ideology took material form in Soviet housing architecture and how such "ideology become matter" also surfaced in people's imaginative and projective inner feelings.

For the Cuban context, Martin Holbraad (2018) has characterized the Cuban Revolution as a totalizing, state-orchestrated political process that flows deep into the minutiae of citizens' lives and, therefore, must be understood as a form of infrastructure. He terms "revolution as infrastructure" the process that forms persons in their everyday encounters with state structures and institutions as these invade people's private spheres. Since the successful Revolution in 1959, the Cuban state has drawn its legitimacy from the promise of an egalitarian infrastructure that guarantees services such as housing, health care, and education for the entire population (Eckstein 1994; De la Fuente 2001; Farber 2011). Cubans were expected to actively participate in the continuous process of socialist modernization and, consequently, the collective creation of a new society, for, as Che Guevara wrote, "to build communism, a new man must be created simultaneously with the material base" (1971, 343). This envisioned construction of material base and social superstructure coincided, for example, in the establishment of the microbrigades campaign initiated by Fidel Castro in his annual speech on July 26, 1970.[7] Microbrigades were collectives of workers who were responsible for the construction of socialist housing units for themselves and their colleagues and therefore built not only a new form of domestic infrastructure (that was designed according to socialist principles) but also a collective consciousness in the process (Stătică 2019).

Today, Cuban citizens experience the state as a paternalistic state that provides them with subsidized food, employment, health care, education, information, culture, and entertainment. Until under Raúl Castro's tenure when a private housing market was legalized, the powerful Instituto Nacional de la Vivienda virtually controlled all their living arrangements as housing was distributed centrally by state authorities. During the so-called Energy Revolution of 2006, it even distributed new and more energy-efficient household appliances such as light bulbs, air conditioners, cookers, and refrigerators. In addition, the Cuban state from early on has sought to incorporate all kinds of civic initiatives into its structures by enrolling citizens in various mass organizations such as the neighborhood Committees for the Defense of the Revolution, or worker, women, student, and peasant associations. For Holbraad (2014, 2018), whose interlocutors mainly belong to the older generation who still remember the times of pre-1991 state socialist normalcy, this orchestrating of the lives of citizens through state planning is a process of infrastructural penetration through which the Revolution as an all-encompassing motion becomes part of the very fabric of their existence.

However, I argue that my (mostly younger) research participants experience in their daily lives also a state other than the paternalistic state, and one that does not appear in Holbraad's analysis of Cuban state-society relations. This state is the corporate state that is in the hands of high-ranking officials and military elites who, since the economic crisis of the 1990s following the collapse of the Soviet Union, have transformed various former state monopolies in the central-socialist economy into untransparent business enterprises. It is institutionally separate from the redistributive socialist state and encompasses various conglomerates under military command and their loyal managers who have immediate access to state leaders (Gold 2015, 165–69; Thiemann and Mare 2021). Military-led businesses control or participate in virtually all profitable sectors of the country's economy, such as tourism, the domestic foreign exchange market, air transport, mining, biomedicine and overseas medical services, the trade of imported goods, and the export of profitable products such as tobacco and rum, and they do so without any public accountability (Tedesco 2018). As state-backed de facto monopolists, they can suppress potential domestic or external competitors and therefore achieve high profit margins. A prime example of such a company, as we will see in the next chapter, is the state Internet provider ETECSA, which charges extremely high prices even by international standards.

As an extensive investigation by the Cuban independent news platform *El Toque* (2022) revealed, the multisectoral conglomerate Grupo de Administración Empresarial S.A. (GAESA), which belongs to the Fuerzas Armadas Revolucionarias (Revolutionary Armed Forces) of Cuba, is by far the island's most profitable company. Its commercial activities extend to at least eleven countries in businesses that serve to evade U.S. sanctions and move money via offshore companies registered in tax havens such as the Bahamas, Panama, and the British Virgin Islands. GAESA was created in 1995 when, at the height of the Special Period, Fidel Castro granted the armed forces control over trade and investments related to the entry of foreign currency. For twenty-six years until his death on July 1, 2022, GAESA was led by General Luis Alberto Rodríguez López-Calleja, who also was one of the fourteen members of the Political Bureau of the Central Committee of the Communist Party, advisor to President Miguel Díaz-Canel, and son-in-law of Raúl Castro.

With the emergence of the corporate state after the dissolution of the socialist bloc, one could say that the postsocialist transformation in Cuba has already occurred in economic (although not in political) terms (LeoGrande 2023). My research participants encounter this corporate state primarily as paying customers of the overpriced products of state-run businesses (Raúl Castro's economic reforms, when Cubans were finally allowed to use cell phones, enter hotels, rent cars, and purchase electronic devices, not only granted Cubans new freedoms but also greatly expanded GAESA's customer base). The corporate state further impacts their lives by limiting their opportunities for economic activity, as the state grants licenses and economic space only to small Cuban businesses that do not clash with its established monopolies. In addition, GAESA controls part of the inflow of remittances that many Cubans rely on since money transfers to the island arrive not in foreign currencies but in the electronic currency MLC, which can be spent only in retail stores that are also part of the network of companies centralized and managed by the conglomerate. Finally, my interlocutors also take note of the string of new luxury hotels the corporate state is building all over Havana, changing the face of the city while the rest of its built environment and infrastructure is left to decay.

The current Cuban system thus simultaneously produces infrastructural inclusions *and* exclusions that shape the lives of its citizens. The paternalistic state continues to regulate the economy in such a way that citizens are still largely expected to work for it, buy from it, sell to it, respond to the demands of its omnipresent apparatus, and make sacrifices for the sake of the state-orchestrated revolution, for example, by working for token payments in the near-worthless state currency (Thiemann and Mare 2021). This paternalistic state has simultaneously withdrawn from many of its former redistributive functions and provides fewer and fewer public services, creating new forms of exclusion that the Cuban Revolution originally sought to eliminate. The corporate state, on the other hand, actively excludes those who do not have access to hard currency because they cannot shop at MLC stores or afford constant Internet access with their peso state salaries. The disastrous state of much of the country's physical infrastructure, with its crumbling buildings and chronic blackouts, is experienced especially by the more vulnerable sections of the population as a form of infrastructural violence. Cuban writer and self-described "ruinologist" José Ponte (2005, 2007) even sees Havana's collapsing urban landscape as symbolic of the country's social decay. As long as residents live in and among ruins, Ponte states, they will not be able to imagine and effect political change.

People as Infrastructure

Anthropology and science and technology studies literature have long emphasized the embedded nature of infrastructures, recognizing that they are defined not solely by their material structures but also by the relationships between human actors, their intentions, activities, desires, and networks (Star and Ruhleder 1996; Larkin 2013; Anand, Gupta, and Appel 2018; Winthereik and Wahlberg 2022). As Harvey, Jensen, and Morita (2016, 5) point out, infrastructures establish, structure, and transform social relations. This can occur through planned activities, which involve intentional and purposeful engineering, as well as through unplanned activities that emerge unintentionally. Therefore, infrastructures are best understood as doubly relational. They exhibit internal multiplicity, meaning they consist of various interconnected elements, and they possess connective capacities that extend outward, influencing and connecting with broader social, political, and economic contexts. In a similar vein, ethnographic research on infrastructure has shed light on how specific local social networks can take on infrastructural characteristics when technical infrastructures fail to adequately meet people's material needs. This occurs in situations where marginalized groups are systematically excluded from public utilities due to racist, classist, or neoliberal austerity policies. It can also arise when the state, due to its standardizing gaze and the embedded classifications within its infrastructures, fails to recognize and address the needs of certain citizens (Scott 1998), or when formerly inclusive socialist states struggle to maintain the services they once provided. In such contexts, social infrastructures that are sustained by relationships between people often come to compensate for the exclusions or dysfunctionalities produced by the official infrastructures by manipulating or supplanting them, for example when people illicitly tap into water lines or electric grids, collaboratively improvise socioeconomic links with one another, create private financial networks, or build alternative waste infrastructures (Simone 2004; Elyachar 2010; Schwenkel 2015; Fredericks 2018; Chalfin 2023).

In the Cuban context, the generative potentials of such flexible and provisional people-embedded infrastructures can hardly be overestimated. Virtually the entire population depends in one way or the other on such support systems to survive in the country's scarcity economy. Entrepreneurial individuals (so-called *mulas*) have established privately operated import businesses that bring all kinds of unavailable but sought-after goods, such as electronic devices, clothing, and spare car parts, into the country—in suitcases and via regular commercial flights (Cearns 2019). Extensive gray and black markets have emerged facilitating private trade in a wide array of goods including food, construction materials, and foreign currency. The products and services circulating in these markets often also stem from the "informal privatization" of public goods, in which state employees utilize state assets such as vehicles, machines, and Internet connections to generate additional private income and use or sell whatever they can steal from their workplaces to make ends meet. These popular economic practices, in which citizens organize themselves to create structures autonomous from the state, are, of course, not a new phenomenon. Barter, private appropriation of state resources, and underground production were a central means for the Cuban people to cope with the economic dislocations of the Special Period and have since evolved into methods for mitigating both the inefficiencies of the paternalistic state's centralized production and distribution systems as well as the exploitative practices of the corporate state's monopolies (that, while highly profitable, are often equally inefficient as they do not have to compete for customers).

However, as I show throughout this book, increasing access to the Internet has created new opportunities for optimizing and expanding such popular economic structures and has led to the emergence of vast digital markets. Simultaneously, the increased accessibility of digital technology has enabled technologically adept Cubans to establish extensive alternative infrastructures for the distribution of online access and content (these were first characterized as "human infrastructures" by Michaelanne Dye; see Dye et al. 2018; Dye 2019). An example of this is the emergence of connection brokers in public Wi-Fi parks who have developed an informal business of redistributing the government's Internet infrastructure. They utilize Wi-Fi range extenders, repeaters, and virtual router software to expand the reach of the Internet, making it available to a larger population at a more affordable price (see chapter 1). Furthermore, media entrepreneurs across the country have collaboratively created *el paquete semanal* (the weekly package), a distribution network for digital data that spans the entire island. The process involves a small group of individuals with privileged government-provided online access downloading content such as film and music files, which is then circulated through physical carriers such as hard drives and USB sticks (chapter 2). In addition, self-taught network administrators have constructed sprawling grassroots computer networks for activities such as multiplayer video games, chatting, and data exchange. An example of this is SNET in Havana, which, at its peak, connected tens of thousands of households through miles of Ethernet cable and thousands of Wi-Fi antennas (chapter 3). These grassroots digital infrastructures not only expand access to the Internet and digital content but also generate parallel economic spaces that enable individuals to liberate themselves from reliance on the state. People can now become independent consumers and producers of content and even establish or promote their own businesses. For a significant part of the Cuban population, these circulation networks have become an important source of income, either directly or indirectly. Simultaneously, a significant portion of the informal economy has migrated to digital platforms, specifically semiprivate chat groups and Telegram channels. These digital realms have become the new hubs for everyday black market activities (chapter 4). Digital platforms have also led to the emergence of new support networks and new forms of civil society engagement where the revolutionary values promoted by the paternalistic state no longer dominate and new moral economies emerge from the bottom up that transgress any socialism–capitalism dichotomy.

Infra(structure) Politics

Many anthropologists invested in the discipline's "infrastructural turn" (Dalakoglou 2016) have sought to conceptualize the world-making capacities of the infrastructural systems they study in political terms. Proposing the motion that "attention to infrastructure offers a welcome reconfiguration of anthropological approaches to the political" at the 2015 meeting of the Manchester Group for Debates in Anthropological Theory (published in Venkatesan et al. 2018), Laura Bear and AbdouMaliq Simone urge us to understand infrastructure as the material, nonhuman ground from which new forms of political relations and new kinds of enactment of collective will may emerge. In her ethnography of garbage infrastructures in Dakar, Senegal, Rosalind Fredericks (2018) sees infrastructures as key sites for negotiated processes of political contestation where elite and disenfranchised citizens alike make claims concerning central ethical and political questions about civic virtue and the shape of citizenship. Disputes over infrastructure thus allow new social and political collectivities to come into being around the use, maintenance, and breakdown of technical systems (Schwenkel 2015). As von Schnitzler (2016) insists, the forms of subversive subaltern technopolitics she has observed among the inhabitants of the Soweto township in post-apartheid South Africa, such as the sabotage or destruction of technical-administrative devices like prepaid water meters, pipes, wires, and official documents and certifications, must be understood as a form of politics with other means, a politics manifested in matter that expands the conceptual and imaginative horizons of how we study and conceive of the political.

While I concur with these authors on the significance of infrastructures in shaping political dynamics, I contend that the infrastructural actions and consequences explored in this book predominantly occur beneath or alongside the realm of political visibility. The Cuban authoritarian political system leaves little room for the overt forms of contestation, protest, or technopolitical intervention that Fredericks and von Schnitzler describe. As a result, my research participants consistently maintained that their engagement in creating, sustaining, or participating in alternative infrastructures outside the state holds no explicit political meaning, agenda, or significance. The distributors of el paquete semanal, the makers of the grassroots computer network SNET, and the administrators of Telegram groups for the swap or sale of scarce consumer goods even actively censor all content circulating through their networks that could potentially be read as a political statement by the Cuban government. As I describe in more detail later in this book, particularly the people involved in el paquete and SNET take pains to convince state authorities that they are providing only harmless entertainment. Taking these claims seriously, while reflecting on the undeniable generative effects of these vernacular infrastructures that have created autonomous spaces for new forms of expression, economic activity, and interpersonal relations, I have come to understand the practices of my research participants as *infrapolitical.*

According to James Scott, who coined the term, infrapolitics refers to a range of actions, gestures, thoughts, and utterances that are not easily recognized as political. These acts and expressions fall short of being perceived as politics either because they go unnoticed within the public sphere or because they exist outside the conventional boundaries of what is typically recognized as political. For instance, they might not reach the threshold of "politics proper," as defined by Chantal Mouffe's concept of politics as "the ever present possibility of antagonism" (2005, 17) or Jacques Rancière's understanding of politics as "an intervention in the visible and the sayable" (2010, 37). Scott

(1990, 183) likens these collective acts and expressions to infrared rays that lie beyond the visible spectrum of political legibility. In Latin American studies, the notion of infrapolitics has recently gained traction as a means of describing a realm beyond politics, encompassing conditions of being that cannot be reduced to political life and that exceed any definition of existence solely based on political determinations (Moreiras 2021). Building on his study of class relations in a Malay rice-growing village, Scott (1985) has developed the concept of infrapolitics into a general theory of how subaltern people who are deprived of access to legitimate channels of expression (and therefore cannot manifest their demands or disagreement in conventional and widely recognized forms of political action or speech) find clandestine and discreet ways to perform their silent resistance against their domination. He evokes a "subterranean world of political conflict" (2012, 113) that leaves little trace in the public eye, but in which many small actions together can have enormous overall consequences. As examples for such forms of quiet opposition through which the dominated surreptitiously counter or minimize their material exploitation, Scott lists theft and pilferage from rulers, poaching, feigned ignorance, shirking or sloppy work, freeloading, squatting, clandestine trade and production for sale, tax evasion, and flight.

As several authors before me have noted, such infrapolitical strategies are what large parts of the Cuban population use for day-to-day survival in the crisis-ridden national economy (e.g., see Eckstein 1994; Sawyer 2005; Allen 2012; Angel et al. 2020). Cubans' economic resistance to their violent infrastructural exclusion (by the corporate state) and inclusion (by the paternalistic state), which I described above, is reflected in people using their low-paying government jobs to procure scarce goods to sell on the black market, rent government vehicles or machinery, or receive bribes or additional payment for services they are actually supposed to provide as part of their regular jobs. Thus, everyday activities such as buying food, finding transportation, and using medical or administrative services are deeply enmeshed in a shadow economy independent of the state. Such illegal activities have become so ingrained in Cuban society that they have become manifested in what Scott calls a "hidden transcript," a pervasive set of offstage speech acts, gestures, and practices with their own language, morals, and customs that contradict the state-promoted revolutionary ideology.

The omnipresent idiom *resolver* (to resolve) captures a pragmatic stance toward scarcity—contriving fixes, cobbling systems, and keeping life moving despite obstacles. While this ethos at times has also been reinforced through government narratives that have asked citizens to be resourceful and self-reliant in moments of scarcity and crisis (and has even been glorified as a mode of resistance to the nation's external enemy), it is commonly understood to mean the ability to navigate legal gray areas and act in parallel with the law with cunning and ingenuity (Sanchez and Adams 2008; Dye 2019). It has collective features because in order to meet their needs and those of their families, people often have to work together, for example, when colleagues turn a blind eye to each other's theft at work or even jointly set up a distribution system for pilfered goods among themselves. As I discuss more fully in chapters 3 and 4, such practices have become widely accepted in people's minds, and resolver as a collaborative effort has thus fostered the bottom-up emergence of new moral economies and notions of popular justice.

Just as these infrapolitical acts avoid to openly challenge the state's politics of distribution, the government itself often refrains from explicitly prosecuting illegal actions like the theft and misappropriation of state assets or engaging in black market activities. By not enforcing its own laws in these cases, the government aims to avoid triggering economic and political tensions that could potentially destabilize the system. Instead, it allows these activities to persist in a state of extralegal ambiguity and primarily intervenes selectively when it deems it necessary. As a result, while the infrapolitical acts of everyday economic resistance carried out by citizens alone may not have the power to overthrow the government, they can significantly undermine and override some of its policies (Thiemann and Mare 2021, 193).

In this book, I intervene in debates surrounding the politics of infrastructure by exploring its role as a medium for the emergence, sustenance, and reinforcement of infrapolitics within societies where citizens are deprived of regular channels for political expression. My aim is to shed light on the concealed and covert forms of infrastructural politics and interventions that remain hidden from both state power and traditional political analysis. Building upon existing studies that primarily focus on the infrapolitical agency of Cuban citizens within the popular underground economy, I broaden the scope to highlight how infrapolitics also manifests within highly technological infrastructures. People engage with or modify these systems to partake in actions and collective endeavors outside the boundaries of official political participation. Rather than openly confronting the government and its institutions, my research participants choose to operate beyond state structures altogether. They construct or participate in vernacular infrastructures as a means of bypassing exclusions produced by the state-provided digital infrastructure, such as exploitative pricing policies. In doing so, they quietly transform these structures, rerouting them toward new purposes. This citizen-led redistribution of access differs from Mouffian antagonism or Ranciérean dissent. It transcends questions of subjectivity, identity, demands, and struggles for power legitimacy that define traditional politics. Instead, it embodies a nomadic inventiveness, creating autonomous spaces through the infrapolitical provisioning of alternative networks and connections.

Just as infrastructure sustains most aspects of contemporary life, infrapolitics provides much of the cultural and structural underpinning of the more visible and headline-grabbing forms of political expression that commonly receive more analytical attention. As Scott emphasizes, infrapolitical forms of resistance can turn into proper politics at any time, as these practices are "continually pressing against the limit of what is permitted onstage, much as a body of water might press against a dam" (1990, 196). As a case in point, the Cuban demonstrations of July 11 would not have taken place without the new forms of autonomous social organization made possible by vernacular digital infrastructures (see chapter 5). Infrapolitics in this sense may represent an anticipation of a future movement or allow actors to preserve, maintain, or perpetuate their agency when the political context precludes any serious chance of achieving tangible political gains (Marche 2012, 14).

Scott's emphasis on agency and his insistence on recognizing neglected political groups as self-formed, rational, and autonomous actors with an inherent authenticity and truth that remains impervious to the dominant ideology of the ruling classes have not been without criticism. As Timothy Mitchell (1990) has argued, this creates a contradiction between Scott's argument that the exercise of power requires what Scott calls a symbolic or ideological dimension and his argument that ideological domination never actually dominates. Therefore, my aim is to broaden our understanding of infrapolitics by shifting the analytical focus from conscious acts of resistance by autonomous actors to the infrapolitical power and potentials of actions that may not be perceived by the actors themselves as oppositional, subversive, or defiant. For example, the individuals who serve as connection brokers in the Wi-Fi parks, the el paquete providers, and many SNET administrators view their alternative distribution networks primarily as a means to generate additional income in an ailing economy with limited opportunities. Users, in turn, engage in these networks primarily for entertainment and a sense of community. However, such free spaces, which are relatively independent of the government and allow minor transgressions of the political order, have the potential to become a breeding ground for oppositional consciousness, as research on social movements and nonviolent resistance has shown (Nepstad 2011). Therefore, I argue that applying the concept of infrapolitics to people-embedded infrastructures enables us to expand our political vocabulary. It allows us to recognize seemingly quotidian activities such as communal practices of redistribution as well as the material and interpersonal labor involved in establishing, maintaining, and repairing vernacular infrastructures, as significant and impactful forms of political action, even if they are not consciously conceived as such even by the people involved in them.

Infrastructuring Publics

As noted by several anthropologists of infrastructure (De Boeck 2012; Collier, Mizes, and Von Schnitzler 2016; Chalfin 2017), infrastructures play a constitutive role in shaping publics. They not only form these publics but also prompt us to reconsider the definitions of the public and private spheres. In Cuba, where all traditional mass media and official communication infrastructures are in the hands of the state and the government seeks to control all civil society initiatives through its institutions, the process that Holbraad calls "revolution as infrastructure" is explicitly designed to prevent the emergence of a (counter) public sphere. By understanding the public sphere through its material and social infrastructures, we can move beyond conventional notions that depict it as an intangible space for the free exchange of political arguments, where citizens can openly debate and question the actions of the state (as famously conceptualized by Jürgen Habermas [1991]). "Infrastructuring publics" (Korn et al. 2019) as a research perspective helps us grasp how both the infrastructural power wielded by the state and the infrapolitical practices of citizens distort and transform the idealized notion of the public sphere as a space for participatory democracy.

Alexander Dukalskis (2017) offers a valuable definition of the authoritarian public sphere, a term that appears oxymoronic within Habermas's normative model. In this concept, the state dominates and manipulates political discourse. The autocratic regime aims to prevent citizens' complaints or demands from entering the public realm, while isolating individuals with critical views about the regime's rule. This is achieved through both "positive" legitimation, presenting the status quo as inevitable and without alternatives, and "negative" repression. Dukalskis's model also recognizes the role of state infrastructure in constructing and maintaining an illiberal public sphere. It identifies five infrastructural dimensions: traditional media, schools, political parties, legislation, and online control. These dimensions work together to shape reality by disseminating messages that legitimize the regime, shaping human capacity through education and direction, and regulating cyberspace. In such a restricted public sphere, citizens are constrained to adhere to what

Scott (1990) refers to as the "public transcript" of politically acceptable discourse. Any resistance or opposition is confined to the realm of the "hidden transcript" of infrapolitical utterances, acts, or gestures.

Dukalskis further identifies three social sites that hold the potential to challenge the dominant state narrative and allow for the emergence of new and potentially critical discourses. These sites include the shadow economy, where citizens can attain greater economic independence; independent journalism, which exposes people to alternative viewpoints; and social media, where individuals can engage in relatively unrestricted political discussions. In the context of Cuba, all these three sites are relevant and have experienced significant expansion due to the growing access to the Internet among citizens. However, while digital technology has empowered individuals to navigate within repressive structures, it has not necessarily undermined the parameters of state control.

The Internet and Political Transformation

The influence of networked technologies on authoritarian regimes has ignited significant discussions in recent years, particularly in the wake of the Arab Spring uprisings during the early 2010s, which occurred in multiple countries, including Tunisia, Egypt, Libya, Yemen, Bahrain, and Syria. It is worth noting that the Arab Spring not only directly affected the Cuban government's policies regarding Internet access, as I explore in the upcoming chapter, but also served as a crucial point of comparison and reference for commentators analyzing the political ramifications of Cuba's momentous July 11 protests. In numerous conversations surrounding digital media and political transformation subsequent to the Arab Spring, a prevailing sense of celebration emerged, highlighting the Internet's capacity to enable decentralized organization, mobilize movements for freedom and accountability, and offer alternative avenues for information and communication beyond the reach of state censorship and control.

However, the notions of "liberation technologies" (Plattner and Diamond 2012), which empower "networks of outrage and hope" (Castells 2015), and bear the potential for "connective action" (Bennett and Segerberg 2013), that might eventually lead to a "Revolution 2.0" (Ghonim 2012), have faced valid criticisms for their technodeterministic outlook. These perspectives have been criticized for their exaggerated emphasis on the spontaneous, unorganized actions of supposedly leaderless "crowds," which, for example in the case of Egypt, disregarded the agency and significant mobilization efforts of organized actors on the ground, such as labor unions (Aouragh 2015). Miriyam Aouragh argues that these narratives of new media and its impact on political contesta-tion not only perpetuate Orientalist stereotypes that portray Arab peoples as needing Western technology to achieve modernity but also reinforce neolib-eral notions of the Internet as a universal, open, and free space that are shaped by U.S. interests and fail to challenge the imperialist power held by major U.S. tech corporations over the underlying technology.

Another influential critique of such cyberoptimism was raised by Evgeny Morozov (2011), who dismisses the idea that digital technology is inherently emancipatory. He believes that online activism (which he derides as "slack-tivism") distracts people from "real" activism and has zero political or social impact. He further highlights the power of the Internet as a tool in the hands of authoritarian regimes to perfect their surveillance, censorship, misinfor-mation, and propaganda. Such pessimistic analyses of digital authoritarianism have increased in recent years in light of the resurgence of dictatorship in the Arab world, increasingly sophisticated online controls in China and Russia, but also various data breaches by big tech companies like Facebook, and the global spread of online deception and vitriol.

In a recent essay addressing the "new structural transformation of the pub-lic sphere" brought about by digital technology, Habermas (2022) expresses concerns about the blurring of boundaries between the private and public realms in the online sphere. He argues that the commercial nature of digital platforms hampers informed exchanges between citizens and poses a threat to the functioning of the democratic public sphere. He highlights the emotional manipulation of users and the potential for political influence as risks associ-ated with the commercialization of social media. Habermas emphasizes that the decentralized and largely unregulated nature of social media platforms has removed traditional filters that previously guided communication in the democratically constituted public sphere. Gatekeepers such as journalists and editors, who formerly directed the flow of communication and determined the reliability and significance of information, have been sidelined. This shift has allowed audiences to assume the role of authors themselves, leading to the emergence of a semipublic, tribalistic, fragmented, and self-centered form of communication. According to Habermas, this transformation distorts the inclusive nature of the political public sphere and undermines the deliberative process of opinion and will formation. He argues that when the infrastructure of the public sphere fails to direct citizens' attention to relevant issues and no longer ensures the formation of competing public opinions that have been qualitatively filtered, the democratic system as a whole is compromised.

Much scholarly analysis of the political impact of networked digital tech-nologies thus still vacillates between glorifying their power to expand the political, social, and economic freedoms of citizens of authoritarian regimes and warning of their potential to enhance the control of autocracies or even undermine deliberative democracy. With this book I hope to show that engaged, long-term ethnographic research can provide a much more nuanced picture of both the power and limits of digital technologies in the hands of citizens of authoritarian states than such grand theories can. Accounts of the causal role of the Internet in political transformations are often projections distorted by a number of analytic biases. The "liberation technology" debate (that unsurprisingly emerged at Stanford University) is steeped in technosolutionist "Californian ideology" (Barbrook and Cameron 1996). The fact that many evangelists of the Arab Spring social media revolutions do not speak Arabic and followed the events only online clearly skewed their understanding of what was happening on the ground and led to technodeterministic analyses that confused the medium and message of political activism (Aouragh 2015, 259). Conversely, pessimistic narratives of digital authoritarianism, exemplified by Morozov's work, tend to underestimate the ecological impacts of networked technology that can weaken the legitimacy of authoritarian regimes. These accounts remain confined to an online-centric analysis of political activism's decline and fail to recognize the interconnectedness and symbiotic relationship between online and offline forms of political organization. Meanwhile, normative conceptions of the public sphere rooted in Habermasian theories overlook the political potentials of horizontal, many-to-many communication, as they primarily envision democracy relying on a professional class of gatekeepers responsible for organizing deliberative discourse.

As I detail throughout this book, digital technologies and increasing access to social media platforms in Cuba have not created fragmented micropublics (as feared by Habermas) but established connections between previously isolated individuals who can now share common interests or concerns, thus strengthening the "weak ties" that can bridge disparate societal groups. As highlighted by Zeynep Tufekci (2017), in an emerging networked public sphere enabled by social media, nonpolitical affordances can hold significant political power. Autocratic regimes find it challenging to censor a large number of users who simply wish to connect with each other or shut down a social media site that primarily serves as a platform for sharing entertaining videos rather than organizing protests. In this vein, the vernacular offline infrastructures and emerging online spaces of socialization that I describe in this book acquire important infrapolitical qualities. The people involved in these spaces, even without explicitly discussing political matters, formulating demands, or engaging in deliberative discourse, experience the power of successful self-organization outside of state structures. For instance, el paquete semanal serves as an autonomous infrastructure that enables the nationwide dissemination of content, contributing to citizens' economic independence from the state. Cuban small business owners, independent magazine publishers, and social media influencers now utilize el paquete as a distribution network for their self-produced content, which was previously absent from state-controlled mass media. In a similar fashion, grassroots computer networks like SNET have brought technology aficionados everywhere on the island together via the collaborative creation of network infrastructures.

While critical discourse is restricted on these alternative networks because the Cuban state has partially outsourced censorship to their administrators, Cubans have much more leeway on commercial global social media platforms, which escape the government's control. Semiprivate Facebook and WhatsApp groups and Telegram channels (which emerged primarily after the advent of mobile Internet) not only have given rise to an expansive online black market but also have fostered the development of new support networks and novel avenues for civic engagement. These networked spaces break down what Tufekci (2017, 26–27) calls "pluralistic ignorance" (the belief that no one shares one's views when actually everyone has been collectively silenced) and make it harder for authorities to stifle or control public discourse. A new generation of young social media personalities and digital activists but also many ordinary Cubans (often using Facebook livestreams) have begun to push the boundaries of the "public transcript." They have taken to social media to demand accountability and pressure the government to live up to its commitments. These platforms further enabled several smaller civil society campaigns that advocated for policy changes on "softer" issues such as animal and LGBTIQ+ rights that are compatible with the ideals of the Cuban Revolution and therefore do not threaten the socialist social and political order. They also have facilitated some protests by artists against government attempts to restrict the country's cultural sphere. These protests, while small, were particularly relevant because, as Sujatha Fernandes (2006) maintains, the cultural sector on the island has long functioned as a surrogate for the public sphere.

Yet Cuba's digital revolution also has its limits. While the borderless nature of networked communication flows has meant that the state has forever lost its information hegemony, it also makes Cuba's emerging networked public sphere much more vulnerable to outside interference, as demonstrated by the online propaganda and misinformation campaigns (from the United States and the ultra-right Cuban diaspora) surrounding the July 11 events (see chapter 5). Most importantly, what made the protests fail was not so much networked but rather "old-school" authoritarianism. The state used its control over the traditional mass media radio and TV (which remain the primary sources of political information for most Cubans) to denounce protesters and mobilize loyalists. It made an example of hundreds of demonstrators, some of whom were sentenced to long prison terms, while it drove most of the few known opposition figures into exile. The protests further petered out for reasons that Tufekci (2017, 77), in her analysis of the power and fragility of networked protest movements, has termed "tactical freeze." While social media facilitated the spontaneous formation of nationwide demonstrations without leadership and prior building of formal collective capacities, the lack of a central organizing entity also meant there was no clearly formulated set of demands, no tools or culture for collective decision-making to adjust tactics or negotiate with the regime, and no strategy for continued long-term action.

Post–July 11, another spillover of online discontent onto the streets was out of sight and the potential for open resistance seemed exhausted for the time being. Young Cubans therefore rather used the transnational networks they had established online to arrange their way into exile, reconnecting with friends and family abroad and sharing tips on organizing emigration through social media. In this regard, Cubans' growing connectivity effectively facilitated a mass exodus that relieved the political pressure and freed the government from a forming opposition. In the long term, however, this endangers the future of the country, as it threatens to lose an entire generation of smart and aspiring people.

Multimodal Anthropology

Current discourses on multimodal approaches in anthropology recognize media production as central to the everyday lives of anthropologists and their interlocutors, who now have access to the same means of representation and often share the same (if not greater) technological capabilities (Favero and Theunissen 2018). Studying the increasingly media-saturated worlds of participants in ethnographic research inevitably generates data in various media, such as voice messages, images, social media posts, chat logs, and videos, that are predisposed to new forms of representation that challenge the primacy of the textual within academic scholarship. The new possibilities for research and dissemination of knowledge in the various media formats that digital technology offers allow us to capture the increasingly complex relations we form with our interlocutors and engage and collaborate with them along media forms that they find relevant to their lives (Collins, Durington, and Gill 2017; Dattatreyan and Marrero-Guillamón 2019; Westmoreland 2022).

One of the key objectives of this project is to actively engage with the alternative media landscape in Cuba, not merely as an external observer but to foster new forms of collaborative work and experimental approaches to producing and communicating anthropological knowledge. Just as my interlocutors have harnessed the potential of digital technology to create vernacular infrastructures and build new communities, I have sought to leverage these technologies to blur the boundaries between researcher and researched, exploring novel paths for shared anthropology. Throughout my research, I have collaborated with various individuals and collectives who have a stake in the Cuban media sphere, including programmers, network administrators, social media influencers, science fiction writers, graphic designers, VFX artists, and members of a hacker/makerspace. One particularly significant collaboration I have entered into is with Cuban artist Nestor Siré. Nestor's artistic practice revolves around the unique characteristics of Cuban media culture. Growing up in a family involved in small-scale media entrepreneurship (his grandfather and uncle rented out paperback novels and VHS cassettes for a living), Nestor draws inspiration from the ingenious methods his fellow citizens employ to distribute media, information, and goods. He was also a member of SNET and has formed a long-standing partnership with one of the matrices that compile el paquete, contributing a curated monthly arts folder that serves as a digital exhibition space and features documentaries and news from the art world. Since our initial meeting in early 2020, Nestor and I have developed a highly productive research-creation practice. This collaboration has resulted in several exhibition projects, which I discuss in this book, as well as multimodal publications (Köhn and Siré 2022a, 2022b, 2022c, 2023).

By immersing myself in the everyday lives of Cuban media practitioners and establishing collaborative relationships with them as epistemic partners (Holmes and Marcus 2008), I gained invaluable insights into the intricate economic, social, and political frameworks that shape their experiences. The fields that I entered often constituted what George Marcus (2010) calls para-sites that were populated by epistemic communities of highly skilled technology experts, who oftentimes shared with me an interest in documenting their community's history and practices. Through working closely with them, I directly encountered the challenges they faced, such as acquiring scarce technical equipment or materials, navigating restrictions on Internet access, and interacting with Cuban state representatives. These experiences yielded profound fieldwork data that revealed the opportunities, limitations, and power dynamics my interlocutors had to negotiate as both citizens and entrepreneurs within the constrained private sector. In many ways, I actively participated in the production processes that formed the subject of my study. This involvement entailed bringing technological devices into the country that were not readily available, assisting in organizing workshops, and cocreating projects alongside research participants. Frequently, realizing these projects required seeking additional local collaborators with the necessary technological expertise, which, in turn, expanded my network of interlocutors.

Multimodal methods do not inherently embody a collaborative approach, and it is crucial to recognize that collaboration itself can inadvertently perpetuate neocolonial power dynamics, especially if it is used merely as a token gesture for Western audiences (Lea and Povinelli 2018). To engage in truly horizontal modes of collaboration, guided by an ethics and aesthetics of accountability (Ginsburg 2018), it is necessary to navigate the complexities of negotiating divergent interests and conflicting goals among all parties involved (Nayyar and Kazubowski-Houston 2020). As I detail in the chapter interludes that describe the production process of each work, I have sought to live up to these ethical demands by sharing authorship and discursive authority, involving and crediting participants as professionals in their respective fields (and not as token collaborators), being transparent about the processes that led to aesthetic and conceptual decisions, and redistributing all production funding I was able to acquire.

These collaborative endeavors were also motivated by a desire to engage a more diverse public that extends beyond the usual reach of anthropological representations. While some projects have been exhibited in international art institutions, others, such as our documentary video game *PakeTown*, were primarily created with the intention of resonating with local Cuban audiences. Each project has been carefully crafted to be meaningful and relevant to the community being studied, and in every instance we aimed to present the work within Cuba itself. This involved drawing on the local alternative vernacular distribution infrastructures or collaborating with state institutions like the Joven Clubs de Computación (Youth Computing Clubs). However, these endeavors necessitated intricate negotiations with local authorities, with outcomes varying in terms of success. These negotiations provided valuable insights into the limits of permissible expression and representation within the authoritarian Cuban public sphere as well as the internal hierarchies and decision-making processes within Cuban state institutions. Throughout these experiences, I observed my collaborators employing a range of strategies, from self-censorship to skillfully advocating for their interests, as they navigated confrontations with these institutions.

In a significant contribution, Stephanie Takaragawa and colleagues (2019) have cautioned against assuming that multimodal approaches in anthropology are inherently liberatory. They highlight that the digital technologies and infrastructures enabling these approaches can inadvertently perpetuate power hierarchies by either indulging in technofetishism or masking neocolonial forms of exploitation. This observation resonates with the complex relationship my Cuban collaborators had with the products and services of technocapitalist corporations, which often remained inaccessible to them. On one hand, my collaborators yearned for the seemingly frictionless access to technology and digital infrastructure available in other parts of the world. On the other hand, they took pride in their ingenuity in developing low-tech DIY solutions as a means to compensate for their exclusion. While many of the participants in my research did not consciously perceive their alternative technological infrastructures, born out of necessity, as a form of resistance against exploitative technocapitalism, members of the Copincha hackerspace engaged in extensive discussions about the positioning of their practices within the broader global power hierarchies.

Whenever I participated in workshops at Copincha or presentations organized by Havana's vibrant community of science fiction writers, readers, and scholars, I was privileged to witness engaging debates revolving around distinctly Cuban technofutures. These events often attracted individuals with a shared interest in technology, leading to overlapping conversations among various communities. The multimodal projects that accompany this book are implicitly or explicitly informed by these discussions. They not only reflect the particularities of Cuban media realities but also playfully explore the question of whether and how the vernacular infrastructures Cubans have collectively created can provide us with a critical view of the homogenized global Internet. The works invite audiences to see these Cuban *inventos* as counteruses of technology or even as viable alternatives to the monopolistic capitalist, consumerist digital infrastructures controlled by a small number of large tech companies.

Outline of the Book

With the opening of public Wi-Fi hotspots in 2015 the Internet came to be distinctively materialized in Cuban public space. In chapter 1 I examine the experiences of novice Cuban users with this public Internet infrastructure, their motivations for spending time online, and the significance it holds for them. I further look at how the state-provided Wi-Fi became redistributed by a clandestine network of informal connection brokers who use Wi-Fi repeaters and virtual router software to extend the signal from the parks to other parts of the city and share their accounts between multiple concurrent users. While this tinkering with the state Internet architecture presents lucrative informal business opportunities, it also makes online access more affordable, efficient, and accessible to a wider range of users.

The chapter is placed in dialogue with *Connectify/Free_Wi-Fi [poesía]* (2020–21), a project by Nestor Siré (in which I was mainly involved as a *mula* who brought the necessary technological components to the island). Nestor set up rigged Wi-Fi modules in public areas in Havana and Miami that display lines of poetry on users' devices instead of the names of available Wi-Fi networks. The featured poems were written by young Cuban poets, some of them living on the island, some of them in the United States. Just as the introduction of Wi-Fi hotspots in public parks represented the first opportunity for many Cubans to reconnect with family members who had emigrated to the United States, these bilingual micropoems are an attempt to engage in transnational communication and bridge one of the most politicized and ideologized borders in the world.

Chapter 2 considers el paquete semanal, Cuba's large-scale offline media distribution network that relies on hard drives and USB sticks. This network is utilized by citizens to compensate for their limited access to international media and the Internet by physically distributing various forms of media content. The chapter contextualizes this phenomenon within the historical framework of earlier informal media circulation networks that emerged as early as the 1970s, initially involving books and later expanding to include VHS cassettes and DVDs. It also examines the profound societal and political implications stemming from the erosion of the former state media monopoly caused by el paquete. I explore how this network grants access to alternative content, fosters a logic of consumerism that contradicts the state-promoted socialist values, and gives rise to a parallel economic space wherein independent business models have emerged.

Another outcome of this research is *PakeTown* (2021), a documentary mobile phone video game about the history of alternative Cuban media distribution I produced in collaboration with Nestor Siré and the Havana-based independent software studio ConWiro. Spanning five decades, the game invites players to become entrepreneurs in the informal Cuban media sector in the format of a business simulation game.

Chapter 3 explores the extensive grassroots computer networks that Cubans have established to compensate for their limited Internet access. Specifically, it focuses on the evolution of SNET, the network based in Havana, examining its rise, decline, and subsequent revival. When the government introduced new Wi-Fi regulations in 2019 that essentially rendered SNET illegal, its members initiated a successful political intervention that resulted in its integration into the national intranet, connecting it with the state-sponsored youth computing clubs. While the fusion with the state-run Tinored led to the loss of SNET's autonomy, anonymity, and decentralized structure, the fight for its survival by its members conveyed compelling notions of successful and independent self-organization. This struggle foreshadowed subsequent protests that had a significant impact on Cuban politics.

In collaboration with SNET administrators, Nestor and I developed an interactive installation called *Fragile Connections* (2022). This installation replicates the technological setup of an SNET network node and serves as a fully operational local area network (LAN). It enables audiences to connect with their mobile devices, explore various sections of the network, and access research materials and documents concerning SNET's history and development.

Chapter 4 delves into the response of Havana's inhabitants during the COVID-19 pandemic, exploring how they turned to semipublic group chats on messaging applications like WhatsApp and Telegram to meet their needs for essential items such as food, hygiene products, medication, and more. This grassroots development of digital sharing platforms, after the introduction of 3G mobile Internet service by the government in December 2018, has fueled the growth of a notable online black market. Simultaneously, it has also nurtured the formation of new support infrastructures and forms of civil society engagement. The interactions within these messaging apps are shaped by conflicting value systems, where solidarity, sharing, and market-mediated exchanges coexist and require constant negotiation.

The site-specific video installation *Basic Necessities* (2021), produced by Nestor and me for the Photographers' Gallery London, provides a visual record of the functioning and aesthetics of the Cuban digital black market via four of the most active Telegram groups, documenting the interactions between some 300,000 users. The work further highlights an eclectic genre of product photography posted by black market sellers—a spontaneous mixture of appropriated stock photography and casual snapshots—that simultaneously offers inadvertent insights into Cuban everyday life. The online extension of the project allows visitors to search live through hundreds of Telegram groups using a digital database. Its content was collected by bots installed in more than four hundred Cuban group chats populated by more than 700,000 users.

In chapter 5, I look at the role of the island's "digital millennials" as active participants in an emerging networked public sphere. As tech-savvy digital entrepreneurs, influencers, artists, musicians, or independent online journalists, they experiment with social media as a means for generating income, mobilizing transnational support networks, building personal brands, and promoting social, political, or economic change. For some of the young participants in my research, this journey led them to become part of the Cuban opposition movement, which reached a historic climax with the social-media-coordinated July 11 protests. Others found themselves in the Cuban diaspora, leaving for places like Miami or Madrid.

The ethnographic documentary *Dinamita* (2022) that I made with Paola Calvo follows the trajectories of Dina (DinaStars) and Adriano (ComePizza), two prominent YouTubers who represent the first generation of content creators in Cuba. Through a combination of our on-camera observations and their self-posted videos, the documentary unfolds over a three-year period, capturing their ascent from fledgling creators to nationally recognized influencers until each of them makes a life-changing decision.

In chapter 6, I revisit the strategies of collective infrastructuring, repair, and self-organization outside of the state that I have described throughout the book and seek to understand them as practices of future making within a failed socialist utopia. Introducing the ethnographic case of Havana's first DIY makerspace Copincha, I investigate how its members, who describe themselves as "citizens of a broken world," have mobilized ideas from cyberpunk and solarpunk science fiction for making sense of Cuban contemporary reality as well as for imagining alternative futures. I discuss the collective's efforts as endeavors to build resilient futures not only for late socialist Cuba but for all of humanity in the face of the imminent climate crisis.

The expanded cinema installation *Memoria* (2023), developed in collaboration with Nestor Siré, Cuban science fiction writer Erick J. Mota, and members of the Copincha collective, is a documentary remake of William Gibson's "Johnny Mnemonic" (1981), one of the earliest cyberpunk short stories. The work appropriates Gibson's fictional story of a data trafficker who has crucial information locked in his head and projects it onto the Cuban media reality with its informal networks and human infrastructures of data exchange and distribution. Examining Cuban grassroots data distribution networks as alternatives to the mainstream capitalist consumerist version of the net that we are left with today, the project envisions more decentralized, (net-)neutral, noncommercial iterations of the Internet.

In the conclusion of the book, I delve into the wider theoretical and practical impacts of Cuba's digital transformation. I highlight how the Cuban context enhances our comprehension of the dynamic relationship between the public sphere and social media platforms, the crucial connection between infrastructures and infrapolitical activities, as well as the impact of digitization on economies outside the capitalist paradigm. This exploration will not only broaden our theoretical insights but also offer practical reflections on how digital advancements interact with and influence societies operating under different political and economic systems.

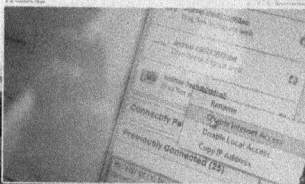

Connectify Hotspot PRO

Settings   Tools   Help

TEAM (8 Clients)

nauta
Servicio de INTERNET
AQUÍ

CHAPTER 1

Wi-Fi

When the Cuban government introduced the first public Wi-Fi hotspots during the summer of 2015, a remarkable transformation occurred in parks and squares across the country. These previously serene and leisurely spaces quickly morphed into vibrant hubs, attracting the attention of crowds of people day and night, all wanting to connect to the Internet. During one of my early, albeit unsuccessful, attempts to access the Wi-Fi at Parque Trillo—the closest hotspot to my apartment in central Havana—Ivan, a computer science student in his mid-twenties with a tall and lanky frame, noticed my struggle. With a quick glance at the recently purchased iPhone 7, which had drained my savings, he cast an approving look my way. Breaking his concentration from his own antiquated laptop, a worn-out device with a scratched surface and faded keyboard keys, Ivan directed my attention to the antenna situated on the opposite side of the square. He proceeded to explain to me the setup of the public hotspots, which typically featured at least two access points to facilitate connection. These access points operated on different channels and transmitted signals in two distinct bandwidths: 2.4 GHz and 5 GHz. With a clarity borne of expertise, Ivan suggested that since most Cuban users possessed older phones incompatible with the 5 GHz standard, I should move closer to the other router to achieve better connection speeds. He went on to elaborate that the 2.4 GHz signal boasted a wider range but was more susceptible to interference due to the prevalence of electronic devices that utilized this frequency band. He even mentioned the potential disruptions caused by neighboring houses' cordless landline phones, painting a vivid picture of the complexities of Cuba's newfound connectivity.

FIGURE 1.1. People in Havana connect to the Internet at various public hotspots, relying on scratch cards or illicit connection brokers who employ Connectify virtual router software to share a single account with multiple clients.

Ivan was among those individuals who frequented the park not to merely browse the Internet but to manipulate and redistribute the signal, turning it into a means of livelihood. In the vicinity of these hotspots, an alternative economy had emerged, where resourceful entrepreneurs offered solutions to the various challenges Cuban netizens faced with their Internet access. Drawing upon interviews and extensive fieldwork conducted primarily in 2017 and 2018, this chapter probes the ways in which Cubans appropriated the state-provided Internet infrastructure, making it their own. It provides an in-depth exploration of the strategies employed by Cuban Internet users to navigate the rationed and costly nature of their access, delving into the reasons behind their online presence and the distinct subjectivities that emerged as they creatively tackled the limitations imposed by this infrastructure.

The exorbitant prices imposed by the state provider, ETECSA, served as a catalyst for IT-proficient individuals like Ivan to repurpose the official infrastructure to suit their own needs. This human infrastructure of informal connection brokers operates parallel to the state-provided network, redirecting the Wi-Fi signal from parks to various locations across the city. I investigate how this appropriation of a state resource, which was (at least initially) scarce due to intentional constraints, constitutes an infrapolitical strategy, allowing Cubans to circumvent the exploitative economic practices of a state monopolist while simultaneously serving as a profitable venture in the extralegal underground economy. Furthermore, this alternative service has been embraced by numerous Cuban netizens as it offers more affordable access to the Internet, making it more accessible to a broader population.

However, prior to Internet access becoming a lucrative revenue stream for the corporate state, the Cuban Internet landscape had been shaped by the domestic priorities of the paternalistic socialist state as well as external constraints arising from the conflict with the United States. The U.S. government pursued a contradictory policy regarding Cuba's access to information and communication technology (ICT), contributing to the complex web of national and international restrictions that Cuban users had (and still have) to navigate. To commence this chapter, I unpack how both local and external factors have distinctively influenced the Cuban experience of the Internet over the two decades preceding the establishment of public Wi-Fi hotspots. Through this exploration, I demonstrate how the limitations imposed on Internet access have consistently attracted entrepreneurial citizens who recognized business opportunities in providing illicit means of access.

A (Pre)history of the Cuban Internet

The integration of Cuba into the global Internet architecture has been shaped by the unique economic, political, and social context that has defined the country since the Revolution. Despite being just ninety miles away from Florida, Cuba has been disconnected from the undersea cables that form the communications infrastructure linking Latin America to the United States because of the U.S. economic embargo against the country. Prior to 1959, Cuba boasted the highest telephone density among Latin American nations (although the majority of landline telephones were concentrated in Havana). However, the revolutionary government did not prioritize significant investment in expanding horizontal telecommunications between its citizens, deeming it an unnecessary luxury. The economic crisis of the Special Period further exacerbated the situation, as it cut off Cuba's supply of equipment and spare parts from the former socialist bloc, leading to a significant decline in the quality of telephone service (Press 1996). Regarding international telephone traffic, the most crucial connection for Cuba in prerevolutionary times had been with the United States. In 1921, the American company AT&T laid the first submarine cable between Florida and Havana. Following the Revolution, the U.S. trade embargo allowed AT&T to maintain bilateral telephone operations between the two countries, but it prohibited network modernization. Consequently, over the years the telephone network in Cuba became almost dysfunctional, requiring long-distance calls from the United States to be routed through third parties using satellite uplinks. Due to the sanctions, AT&T also was unable to make direct payments to Cuba for their share of bilateral telephone traffic. Instead, these funds were deposited into an escrow account in the United States (Hoffmann 2004, 157).

U.S. communications policy toward Cuba underwent a significant change in 1992 with the passage of the Cuban Democracy Act, commonly known as the Torricelli Act. This act tightened the embargo by prohibiting U.S. subsidiaries in other countries from trading with Cuba. Additionally, it targeted countries that provided aid to Cuba by denying them debt reduction or forgiveness. However, alongside these tightened sanctions, the Torricelli Act also introduced some selective exemptions. These exemptions allowed for family remittances, postal services, and limited telecommunications between the United States and Cuba. This approach, known as the "second track," aimed to facilitate the emergence of a civil society in Cuba. It was hoped that this civil society would oppose the government, ultimately leading to a transition to democracy (U.S. Congress 1992, sec. 1705). While the Cuban government criticized the Track II policies, accusing them of ideological subversion, some practical benefits did accrue to the country. The restoration of direct telephone connections to the United States made the Cuban market attractive to foreign investors who were urgently needed to improve the country's deteriorating telecommunications infrastructure. In 1994, the Cuban Ministry of Communication successfully signed a joint venture agreement with the Mexican Domos Group (later joined by Telecom Italia) to establish ETECSA (Empresa de Telecomunicaciones de Cuba), the new telephone company of Cuba. However, Domos Group later sold its stake due to financial difficulties, mainly caused by the Mexican peso crisis. Furthermore, following the passage of the Cuban Liberty and Democratic Solidarity Act (more widely known as the Helms-Burton Act) in 1996, which again reinforced the embargo (U.S. Congress 1996), U.S. pressure led to further complications for Domos Group (López 1999). Eventually, Telecom Italia acquired Domos Group's stake and heavily invested in the transition from analog to digital telecommunications infrastructure in Cuba in the subsequent years.

A planned U.S.-Cuba fiber-optic submarine cable project did not come to fruition due to the Helms-Burton Act because it included a provision that prohibited any investment in the domestic telecommunications network within Cuba. The ambiguous wording of this law hindered any plans that involved connecting a submarine cable between the United States and Cuba to the island's national telephone network. The Helms-Burton Act, along with the U.S. government's rejection of the cable plans, was just the beginning of a series of decisions that demonstrated the inconsistent management of the Track II policies that were aimed at supporting civil society by expanding Cuban citizens' access to ICT. Subsequently, the direct-dial service via satellite between the United States and Cuba was discontinued by the Cuban side in 2000 due to new diplomatic tensions. Cuba protested that the United States had used the money escrowed for long-distance calls as compensation for the damage caused by the Cuban Air Force's shooting down of a plane belonging to Brothers to the Rescue—an exile Cuban activist organization that rescued raft refugees attempting to emigrate from Cuba—which had entered Cuban national airspace (Morley and McGillion 2002, 170).[1]

Since the early 1990s, cellular service in Cuba has been provided through an international joint venture called Cubacel. This joint venture is a collaboration between the Cuban government and the private company Telecomunicaciones Internacionales de Mexico. In 1998, Canadian mining and energy company Sherritt International also became a shareholder in Cubacel. However, Cuba later decided to renationalize its telecommunications sector. In 2003, the

Cuban government bought back shares from Sherritt International, followed by the repurchase of shares from Telecom Italia in 2011. These actions were taken to consolidate all telecommunications services in Cuba under the umbrella of ETECSA, a single, fully state-owned monopoly telecommunications provider (Recio Silva 2013).

While the paternalistic socialist state had never shown much interest in providing private telephone lines to its citizens, it recognized the potential social benefits of computer technology early on. In the 1970s, Cuba made efforts to build a local computer industry with the support of the Soviet Union. Unlike other countries with similar economic development levels, Cuba invested significantly in domestic computer networks that primarily connected government and scientific institutions (Press 2011). In 1987, the first Joven Clubs de Computación were opened in Cuba, driven by the personal initiative of Fidel Castro. These community computer centers aimed to familiarize Cubans with technology. During this era, the policy and regulatory framework emphasized the social use of ICTs for the benefit of the socialist nation (Venegas 2010). Cuba established notable computer networks for professional use, such as Infomed, a health portal connecting doctors, pharmacies, hospitals, and research institutions, providing public access to medical information. Another network was Tinored, the intranet of the Joven Clubs. The development of these intranets reflected Cuba's aspiration for digital sovereignty, a policy goal that remains important today.[2]

Regarding the country's connection to the global network, however, the government remained hesitant for a long time. Intimidated by the U.S. reasoning behind the Cuban Democracy Act's Track II, which suggested that the free flow of information would encourage pluralistic tendencies, strong forces within the Cuban leadership feared the potentially destabilizing effects of connecting to the Internet. According to Bert Hoffmann's (2004) seminal study on the history of the Internet in Cuba, Fidel Castro himself made the ultimate decision to permit access after apparently months of weighing the potential positive and negative implications. Hoffmann identifies three key factors that influenced Castro's decision. First, the worst effects of the social and economic crisis of the Special Period, which characterized the Cuban reality at the time, were slowly subsiding. Second, China signaled its willingness to share its technological and administrative expertise to minimize the potentially dangerous national security implications of ICT. Third, the Cuban government dispelled emerging debates about the need for greater pluralism and an independent civil society, which had surfaced in intellectual forums like the journal *Temas*, making it clear that the opening of the Internet did not signify a broader trend toward political liberalization.[3]

In August 1996, ETECSA signed a contract with the U.S. Sprint Corporation, enabling Cuba to establish a 64 kbps satellite connection with the United States for a monthly fee of $10,000 (Hoffmann 2004, 170). This connection was made possible by the provisions of the Torricelli Act and did not violate the Helms-Burton Act since it did not involve investment in domestic infrastructure. In October of the same year, Cuba finally joined the Internet, becoming the last country in Latin America to do so. However, consistent with the state's vision that the use of ICT should primarily benefit the (socialist) society as a whole, the government imposed restrictions on individual access, officially citing the country's technical and financial limitations as the reason for these constraints. Private Internet connections were not legally available for purchase by Cuban citizens, and access was limited to official institutions and public access centers where usage could be monitored and controlled.

While Fidel Castro initially viewed the Internet in the mid-1990s as a potentially dangerous instrument of ideological subversion, his views began to relax toward the end of the decade. Milena Recio Silva (2013) has traced how Castro came to see the Internet as a propaganda tool that Cuba could utilize. Inspired by the rise of the antiglobalization movement and its use of the Internet for disseminating information and organizing collective action, Castro increasingly recognized the potential of the global web for the Battle of Ideas campaign. This ideological offensive in the early 2000s aimed to conclude the economic adjustments of the Special Period, reconnect Cubans to the ideals of the Revolution, and promote the Cuban model of social development on an international scale. As a result, Internet access was extended to individuals who could help spread Cuba's message, including progovernment intellectuals, artists, and journalists. Cuban state newspapers transitioned to digital platforms and engaged in a cultural and media battle against U.S. imperialism, targeting an international audience. To support this campaign, the Universidad de las Ciencias Informáticas (University of Information Sciences) in Havana was founded in 2002 to train future computer specialists. However, other educational institutions, such as schools, were not prioritized and remained disconnected from the Internet, revealing a rather utilitarian and narrow perspective on the social benefits of Internet access, as Silvia Oramas Pérez (2016, 44) notes.

According to my older research participants, a black market for private Internet connections emerged alongside the expansion of official access in Cuba. Individuals with workplace access to the Internet started engaging in illegal activities, such as importing computers and modems to use their work accounts from home. They then rented out their access to others, creating a parallel source of income in hard currency. Additionally, some legitimate

users, including employees of authorized institutions, joint venture companies, and foreigners residing in Cuba, also rented out their passwords and
configuration data for specific periods, such as during non–business hours. In
some cases, network administrators even set up and sold additional accounts
unlawfully. Due to the exorbitant prices associated with this illegal access,
some illicit users even began subletting their Internet access by the hour to
share the costs or offered services like receiving and sending emails on behalf
of others for a fee.[4]

Internet Access under the Corporate State

Under the economic and social reforms initiated by President Raúl Castro from
2010 onward, Cuba's approach to Internet access, which previously prioritized
controlled and socially beneficial use of ICT based on socialist criteria, began
to shift toward a market logic. The state-owned telecommunications company,
ETECSA, started charging unsubsidized prices for its Internet services when
providing them to other state institutions, and restrictions on individual users'
access started to be gradually lifted. By 2008, private users were allowed to
import computer equipment and cell phones, although modems and Wi-Fi
antennas remained available only through the black market. Private users could
also enter into cell phone contracts with ETECSA, which had to be paid in
CUC to generate hard currency revenue for the state. These reforms expanded
citizens' options for consumption and transformed former recipients of state
redistribution into paying customers of state monopolies.

The increased availability of computers also led to the expansion of informal Internet access providers. For example, my friend Chuli recounted his
experience in 2010 when he set up a Facebook account at one of the informal
casas de conexión (connection houses) that were prevalent in Havana's affluent
neighborhoods at the time. These private Internet cafés typically shared one
user's legitimate account among several computers, with customers paying
an hourly fee to go online. Often, the families of doctors on medical missions
abroad, who were granted home Internet access to stay in touch with their
relatives, ran these establishments. As these businesses operated in a highly
illegal manner, they required new clients to be introduced by existing customers and demanded advance reservations. The connections were typically slow,
as they utilized modems and were shared between three to four users. Chuli
recalled that for the one CUC he paid per hour, he could do little more than
send a few emails, chat with relatives abroad via Yahoo Messenger, or open
his Facebook profile.

Raúl Castro's economic liberalizations also prompted the newly elected Obama administration to ease some sanctions. In 2009, the United States allowed its telecommunications companies to establish fiber and satellite links with Cuba and enter into roaming agreements with Cuban operators. It also permitted U.S. residents to pay for telecommunications, satellite radio, and television services provided to people on the island. A U.S. company received approval to build and operate a new submarine cable that would connect the two countries, but it was unable to reach an agreement with the Cuban government. Meanwhile, the Cuban state had been working on a solution that aimed to ensure more digital sovereignty. In 2008, WikiLeaks disclosed an official document signed two years prior by Cuban and Venezuelan officials, outlining their agreement to construct a fiber-optic cable between the two nations (Assange 2008). Instead of opting for a direct connection through the United States, Cuba thus chose to establish a connection to its socialist ally. Bypassing the United States, however, meant forgoing the much cheaper and faster option of linking to one of the existing international submarine cables surrounding the island. While the newly laid cable finally arrived near Santiago de Cuba in February 2011, there was little information available in the Cuban media regarding its implementation. It was not until July 2012, when Venezuelan authorities unilaterally declared the cable operational, that news of its status emerged (Recio Silva 2013). In January 2013, ETECSA issued a brief statement confirming that the ALBA-1 cable was in a testing phase but cautioned that this did not guarantee immediate improvement in access. Finally, on May 28, 2013, ETECSA announced its first public Internet access service, Nauta.

There is speculation regarding the reasons for the more than two-year delay between the cable's completion and its launch, considering the substantial public investment of $60 to $70 million. One theory, proposed by Freedom House—a U.S. government-funded NGO dedicated to promoting democracy and human rights (noting that their analyses may be biased toward U.S. interests)—suggests that the Cuban authorities might have delayed opening up access influenced by the media coverage of the Arab Spring. This event was portrayed in Western media as a "social media revolution," potentially causing concerns among Cuban authorities (Kelly et al. 2013, 218–19). U.S. officials, in turn, had similar visions about a "Cuban Spring" fueled by digital communications technology. In 2010, USAID (U.S. Agency for International Development) secretly initiated the development of Zunzuneo, a social network and microblogging service. Operating on the basis of cell phone text messages, Zunzuneo specifically targeted Cuban users with the aim of encouraging them to revolt against their government. The service was designed to attract subscribers by offering harmless content such as sports and music, then gradually introducing political messages to foster dissent. According to the Associated Press, which broke the story (Butler, Gillum, and Arce 2014), the service reached its peak with forty thousand users in Cuba, none of whom were aware that they were using a platform provided by the U.S. government. However, in 2012 Zunzuneo abruptly disappeared due to a lack of funding. As noted by Lana Wylie and Lisa Glidden (2013), the idea that the United States could employ communication technology to instigate a revolution in Cuba not only reflected a popular, yet highly deterministic, belief in the link between technology and political change but also represented a manifestation of the paternalistic and imperialistic mindset that often characterizes U.S. policy toward Cuba.

The controversial Zunzuneo project, despite its intentions, ultimately had no significant impact on Cuban civil society. It highlighted once again the absurdity of U.S. policy toward Cuba, which favors clandestine operations instead of establishing a clear legal framework for improving Internet access for Cuban citizens. Despite the efforts to remove restrictions under Track II of the Cuban Democracy Act during the Obama administration, Cuban users continued to face obstacles imposed not by the Cuban government but by U.S.-related factors. While the U.S. allocates millions of dollars to "digital democracy" programs on the island annually (Eaton 2021), the trade embargo still today directs U.S. tech companies to enforce sanctions on their software and services. As a result, Cuban netizens are blocked from accessing popular applications like Zoom, Spotify, and the Apple App Store. They are unable to purchase digital goods and services from U.S. companies, and certain U.S. products, such as the Apple iPad, do not even include Cuba as an option for first-time user registration. Although some Cubans have resorted to using VPNs (virtual private networks) to bypass geoblocking and access free online services, these restrictions heavily impact the Cuban digital entrepreneurs I met throughout my research. These entrepreneurs, often cited in U.S. government reports as potential agents of political change in Cuba, face limitations in promoting their businesses within the country, as they cannot pay for ads on platforms like Facebook or Instagram. Furthermore, conducting business with the large Cuban exile community in Florida remains challenging, if not impossible, due to the restrictions in place.

The Cuban government, in response to its "dictator's dilemma" (Boas 2000), which involves weighing the risks and benefits of transnational information flow, adopted a strategy of tightly controlling the Internet. In 2008, through Resolution 179, the government tasked ETECSA with the responsibility of censoring content that contradicted the "social interest, morals and good customs, as well as . . . the integrity or security of the State" (Ministerio de Justicia, República de Cuba 2008, 1302, my translation). Consequently, the state continues to block websites critical of the regime and independent online journalism platforms. Additionally, it employs a significant number of progovernment online trolls, who are paid to influence online discussions and narratives (Khrustaleva 2021).

However, the primary obstacle to Internet access imposed by the Cuban government is not a technologically advanced "Great Firewall" similar to the one employed in China, which systematically filters out opposition content, but the exorbitant prices set for access. The state, through its ETECSA monopoly, has made the sale of Internet services to citizens a major source of revenue, enabling it to extract foreign currency from the population. As will be illustrated, this situation even affects Cuban exiles who wish to communicate with their families on the island, turning them into customers of the Cuban corporate state.

The First Public Access Points

The first notable advancement in public Internet access provided by ETECSA, following the activation of the new ALBA-1 fiber connection cable, was the establishment of 118 *salas de navegacíon* in 2013. These government-run centers charged users an hourly fee of 0.60 CUC for access to the national intranet, which included services like a national email provider and politically sanitized versions of popular Internet platforms such as EcuRed (similar to Wikipedia) and Reflejos (inspired by WordPress). Access to international email services was available for 1.50 CUC, while connection to the global Internet carried a price tag of 4.50 CUC. Considering the average monthly income at the time was 23.36 CUC (according to ONEI 2015), these costs proved to be a substantial barrier, making the service unaffordable for a significant portion of the population who didn't receive remittances and lacked additional income from the private sector. The United States continued its policy shift toward Cuba's telecommunications sector following the simultaneous declaration by Barack Obama and Raúl Castro on December 17, 2014, announcing the resumption of diplomatic relations between the two nations. Obama emphasized that improving Cubans' access to ICTs and global communication was a major objective of this diplomatic initiative, and he projected the exclusion of certain necessary technologies from the embargo. Shortly after, ETECSA announced its negotiations with U.S. IDT Domestic Telecom to reestablish direct communication links, including voice calls, roaming services, and direct postal service, between the two countries. The Obama administration also permitted collaboration between U.S. tech companies such as Google, Airbnb, and Cisco with Cuban state institutions. Google was even allowed to establish its Global Cache servers within ETECSA data centers, facilitating better-quality access to its services.

For its part, the Cuban government demonstrated a commitment to improving network coverage and attracting more potential customers. In June 2015, a leaked document titled "Estrategia Nacional para el desarollo de la infraestructura de conectividad de Banda Ancha en Cuba" (Ministerio de Communicaciones, República de Cuba 2015) started circulating online. The document outlined a vision for 2020, aiming for 95 percent broadband coverage in urban centers, 90 percent coverage in rural areas, and Internet access for 50 percent of households at a minimum speed of 256 kbit/s for a price not exceeding 5 percent of the average national salary. The Cuban Ministry of Communication did not confirm the authenticity of this document but also did not deny it (Oramas Pérez 2016, 56).

The most significant change for Cuban Internet users occurred when the state-run newspaper *Juventud Rebelde* reported on June 17, 2015 (Guevara 2015), that ETECSA would deploy the first thirty-five nationwide Wi-Fi hotspots in public parks and plazas by the end of the month. Shortly after, ETECSA announced a reduction in the hourly fee for accessing the global Internet at these hotspots, bringing it down to 2 CUC. The number of hotspots across the country increased rapidly from 314 in 2016 to 1,095 in 2020. Additionally, from late 2016, ETECSA started offering home Internet services through ADSL connections, initially limited to Havana's touristy Habana Vieja district. However, due to limitations such as home wiring conditions and unavailability outside select urban areas, this service did not attract high user numbers, connecting only 189,000 users nationwide by May 2021. Until the introduction of a 3G mobile network in December 2018, which provided permanent Internet access to users with smartphones and sufficient financial resources, public Wi-Fi hotspots thus remained the primary access point for most Cubans. These hotspots were particularly significant as they were the locations where the majority of the population had their initial experiences with the Internet.

The Field Site: Parque Trillo

In the following, I provide an ethnographic portrait of one of the central locations in my fieldwork, Parque Trillo. Located in the Cayo Hueso neighborhood of Centro Habana, Parque Trillo is a large urban square that served as a significant site for my research. At the center of the square stands a towering ceiba tree, which holds sacred significance in the Yoruba pantheon, believed to provide a glimpse of deceased family members to those who walk around it. A prominent black statue of Quintín Bandera, an Afro-Cuban general who fought in Cuba's independence wars, stands in the square. Uncomfortable concrete benches are scattered throughout, offering limited seating. This unassuming square, situated close to my residence, served as the hub of my research. In recent years, numerous data copy stores known as *puntos de copia* had emerged in the surrounding area. One such store, Cinemania 4K, features prominently in the following chapter on offline data circulation. Adjacent to the square, atop a large socialist apartment building, the central backbone of SNET, Havana's popular gaming network discussed in chapter 3, resides on the rooftop. Just a few streets away, nestled within the house of its founder Mauri, Havana's inaugural hacker and makerspace Copincha finds its headquarters, setting the stage for chapter 6.

In 2016, when rectangular white Wi-Fi antennas from the Chinese company Huawei were installed on the wooden electricity poles around Parque Trillo, it brought about significant changes in how the local population utilized the park. Arminda, a sixty-two-year-old resident, recalled that after the square had stopped serving as a weekly fruit and food market some years before, hardly any people stayed there because the benches offered little shade. However, when the park was turned into a Wi-Fi hotspot, people flocked to the square at all hours. As soon as the scorching afternoon sun started to set, large crowds gathered to connect to Facebook or engage in video chats with their relatives abroad. During the late-night hours, the park became popular among the group of millennial YouTubers I follow in chapter 5. They would come to upload their latest videos when fewer people were present, ensuring limited competition for the available bandwidth. Arminda, who had no personal experience with the Internet, expressed her annoyance with the noise generated by some of these new visitors, particularly during nighttime, and the litter they left behind.

The uncomfortable user experience in Parque Trillo still reflected the paternalistic socialist state's view that private access to communication media was not a high priority and not something to be facilitated. The access available in the park was paradoxically both mobile and stationary. Like many consumers in developing countries, where older communication infrastructure is often limited or underdeveloped, most Cubans bypassed traditional fixed-line Internet access, jumping directly to mobile Internet. This shift was enabled by the widespread adoption of Wi-Fi-enabled phones and laptops, which are essential for connecting to public hotspots. Since the Cuban government lifted the ban on private ownership of cell phones and computers in 2008, individuals have obtained these devices as material remittances from Cuban emigrants or through purchases on black market digital platforms like Revolico, a topic discussed in chapter 4. However, until the mobile Internet network was introduced in December 2018, these mobile devices required proximity to stationary hotspots to establish a connection. As a result, people often expressed that they "go to" or "have been at" the Internet when referring to their visits to these Wi-Fi parks. Therefore, the Internet provided by the Cuban state was distinctly marked by its connection to physical locations, emphasizing its material embeddedness in specific places.

The impact of the materiality of the Cuban Internet on the urban landscape of Havana, particularly in Parque Trillo, was evident in various ways, influencing spatial, social, and economic arrangements. Depending on the time of day, Internet users gathered closely together near the Wi-Fi antenna to obtain the best signal strength while also seeking shade. Some individuals brought their own chairs to enhance their comfort while browsing or to distance themselves slightly from the bustling crowds. Street traders selling sweets and pastries took advantage of the crowds, moving through the park to find potential customers. Cars drove around the square, with drivers searching for areas where they could receive a strong signal to access the Internet from the privacy of their vehicles. Despite everyone's efforts to speak quietly, it was unavoidable to overhear people's intimate video conversations. Mothers discussed the need to buy a new washing machine with their children abroad, friends shared the latest neighborhood gossip with family members in exile, and lovers separated by migration made jealous remarks about ambiguous Facebook posts. While people generally attempted to respect each other's privacy by discreetly ignoring these conversations, being online in this environment undeniably turned into a public experience.

Online Activities at Parque Trillo

During my interviews conducted in the park between 2017 and 2018, nearly all of the individuals I spoke with expressed their concerns about the high cost of Internet connectivity and the frustratingly slow network speeds, which caused them significant stress. They constantly worried about not being able to accomplish what they desired online. In response to these challenges, they developed what Jonathan Donner refers to as "metered mindsets" (2015, 123). They adopted strategies to optimize their limited online time and focused on only the most important activities. For most of the people I interviewed, the key benefit of their online experience was the newfound ability to stay in touch with family members who live overseas. Given that roughly one-third of Cubans have family in the diaspora, the Internet has bridged a significant gap, reuniting them with relatives like parents, siblings, children, and grandchildren. This connection was long obstructed by the U.S. embargo, which prevented direct phone and mail communication with Cuba until 2015. Moreover, the Cuban government's stance on its expatriate community, often pejoratively called *gusanos* (worms), added to the separation. Yet, with the advent of Wi-Fi hotspots, people could now engage in regular video calls and see and hear the voices and faces of their long-lost family members. Julio, fifty-five, for instance, who visited the park at least twice a week to speak with his sister, vividly recounted an experience where she took him on an impromptu video tour of her home in South Florida during one of their early video calls:

> "Let's go for a walk," she said. "Ah, you have to go, we have to stop talking," I replied. "No, we can keep talking," she insisted. With that, she proceeded to give me a tour of every room in her house, including the garden and the pool. Since she had been abroad for almost twenty-five years, I didn't know her living situation. And then, at last, it was right before my eyes—I could finally see it!

Facebook, in particular, played a significant role in facilitating connections for many of my interlocutors. Its search function allowed them to locate expatriate friends and distant relatives whom they believed they had lost touch with forever. Benito, aged fifty-nine, shared with me his experience of rediscovering one of his closest childhood friends through the platform:

> His name was Orlando, he was of Chinese descent, so we nicknamed him El Chino. He was known for his beautiful singing. In my previous attempts to find people, I was successful, so I decided to search for his name. As I was looking through profiles, I came across one and exclaimed, "Is this the Chino?" I immediately messaged him in the chat, "Chino! Brother! It's me!" He was overjoyed. To our surprise, we discovered that he lived near my sister in Homestead, and he wasted no time in visiting her. He even sent me a plastic jug, the kind that can be frozen, for beer. I told him, "Chino, but I can't drink beer, I'm diabetic," but well, here I have it in the fridge.

Many Cuban users are able to afford to access the Internet only due to the support they receive from friends and family members outside the island who top up their Nauta accounts. This option serves as a means for the Cuban government to acquire much-needed foreign currency. Platforms like Ding, Cuballama, and Suenacuba enable customers outside of Cuba to conveniently recharge Cuban accounts. ETECSA even periodically launches promotions that offer extra bonuses for top-ups made from abroad. As an investigation by the independent news platform *Havana Times* revealed (López Moya 2022), the state is extracting most of the revenues from the company, leaving ETECSA employees struggling to maintain quality service because the company cannot make large investments in the telecommunications infrastructure.

In addition to enabling online communication, the increased contact with family members overseas has also had a significant impact on the amount of remittances received by individuals in Cuba. Regular contact through the Internet has made relatives more aware of the needs of their loved ones on the island. As a result, the level of financial support has often increased. Furthermore, these connections have fostered collaborative investment projects, with members of the Cuban diaspora providing financial assistance for small business ventures or restaurants in Cuba. An example of this is José, aged thirty-nine, who convinced his brother in Miami to assist him in acquiring the necessary equipment for a copy shop he plans to open in Santos Suárez, where his mother resides.

> I studied the area, not a market study, but I did my research. In the vicinity, there are approximately five schools, but there is no one around who offers that kind of business, absolutely no one. The closest option is about ten to twelve blocks away. I can live with this competition since they even offer photography services. I, however, will focus on printing, photocopying, and scanning for now. I will have one computer and one machine. My brother will send it to me from the U.S. because they are very expensive and not of high quality here. People here often sell old items, assuming that you don't have Internet access and can't check on platforms like Amazon or Walmart, so you wouldn't know.

Besides connecting with friends and family in the diaspora, seeking information and entertainment were the other activities my interviewees reported undertaking online, but only when this material could not be obtained offline through Cuba's alternative media distribution networks (see chapter 2). Maykel, for example, described how he downloads software and applications for his phone or PC that he didn't find in the paquete semanal. Sometimes he even downloaded music videos or TV series he couldn't find anywhere else. His example showed that many Cubans used the state-provided Internet, due to its technical limitations and cost, merely as a complementary network to other, informal distribution infrastructures. They only did things online for which there was no offline alternative in place.

The two most common applications I observed people using in the parks were Facebook and imo, a chat application optimized for low bandwidths that allows access to multiple chat platforms. The connectivity experience in these parks was characterized by users having to undertake various infrastructure-related tasks themselves to maximize their limited access and cope with low network bandwidth. They sought out software solutions such as Facebook Lite, a data-saving version of Facebook, or Opera Mini, a simplified web browser. This kind of infrastructural labor often fostered a sense of community as users shared their knowledge and insights with others in the parks. Accessing the Internet became a collective endeavor for many users. For instance, several interlocutors mentioned that they managed a friend or family member's Facebook page, uploading photos and accepting friend requests on their behalf because these individuals couldn't afford a smartphone. Others took on the responsibility of performing online tasks and running errands for acquaintances. Rodrigo, a man in his late fifties, shared how his friend often asked him to research information whenever he knew Rodrigo was going to the park.

He is a person of great cultural knowledge, someone who has experience in international business. However, I don't see him having any command of the Internet. Whenever he needs information from the Internet, he calls me and asks, "Could you please look up something for me? For example, the sugar prices in 2012. How much was a ton of sugar worth back then?" I responded, "But can't you search for that yourself?" He replied, "No, no, no, absolutely not! I can't do it." So, there is this group of individuals, a segment of the population that faces this problem of not knowing how to access the Internet.

As I discuss in the following, those who lacked basic knowledge of the Internet or couldn't afford the official prices could also rely on a parallel infrastructure that had developed in the public spaces of the parks. In Parque Trillo, as well as at other hotspots, an important, if unsung, human element operated within the island's Internet infrastructure, providing customers with both technical support and cheaper connections.

The Park's Connectivity Entrepreneurs

At Parque Trillo, like in almost all the other Wi-Fi parks I visited, there existed an informal network of individuals who engaged in interface work to generate additional income. These connection brokers earned a living by illegally reselling Internet access cards or utilizing Wi-Fi range extenders and virtual router software like Connectify to distribute their accounts among multiple users simultaneously, thereby reducing the cost per user and making access more accessible at a cheaper price. To access the Internet in the parks, users needed to purchase *tarjetas* (scratch cards), which came in the form of onetime cards with a twelve-digit username and password for either one or five hours, or top-up cards for those who had a permanent Nauta account. However, obtaining these cards was not always straightforward. While recharge cards could be bought from private individuals who held a cuentapropista license as *agente de telecomunicaciones* and usually sold them from their homes, disposable cards were exclusively sold through official ETECSA offices. Personal data of buyers were registered at these offices to ensure that Internet usage was not anonymous. However, these stores were often not conveniently located near hotspots and closed early, and long queues commonly formed outside them. To compound the difficulty, ETECSA offices restricted the maximum number of cards to three per buyer per day. Consequently, when an Internet user exhausted their cards, their only option was to obtain a resupply from a *tarjetero*, a person informally reselling cards for 3 or sometimes 2 CUC (in the summer of 2017, ETECSA had reduced the official price per hour to 1 CUC).

In Parque Trillo, each of these tarjeteros had a designated spot in a specific corner of the park. Often, these resellers worked in teams, with one person guarding their territory while the other waited in line at the next ETECSA office. Although this practice was technically illegal, it carried relatively low risk due to the lack of controls and the prevalence of informal resale in various sectors of the economy (see chapter 4). King, a hip-hop musician in his early twenties who sold tarjetas on the side at the park, mentioned that while he had been questioned by the police a few times, they had never gathered enough evidence to fine him. He conducted his sales discreetly: a brief exchange of glances with the customer, followed by a short negotiation, then he opened his hip bag, and the money and cards changed hands.

In other corners of the park, much closer to the actual antennas, another informal business took place. Connectivity entrepreneurs such as Ivan, the connection broker who assisted me in connecting to the hotspot in Parque Trillo for the first time, shared their Nauta accounts for a much cheaper fee than the official rate—prices ranged from half the official rate to 1 CUC for several hours of surfing. Although connecting multiple users through a single account was also illegal and made the connection even slower, this offered a viable and widely available alternative for users with less money to spend on going online. As Marta, a neighborhood literature student, told me, she preferred to pay per connection rather than pay for time spent:

If I'm going all the way out here to do some surfing, I generally prefer to stay online for a longer duration. With one of these guys working in the park, I pay the same amount as I would for an hour of access, but I can stay online as long as I want. Sure, the speed may be slower, but that doesn't bother me. What really stresses me out is constantly seeing my credit minutes disappearing.

The connection brokers were usually identifiable by their position in the center of a small crowd, seated behind a laptop running virtual router software. In the case of older models, they often had an easily recognizable USB Wi-Fi adapter with an attached antenna. They often worked with assistants who would roam the park, bringing them new customers or helping clients connect to their bosses' private hotspot network and enter the password on their devices. While, as Lisa Parks and Nicole Starosielski (2015, 6) note, users in capitalist societies are typically taught to appreciate the conveniences of modern consumer technology without understanding the supporting infrastructures, many of my Cuban interlocutors had amassed significant knowledge about tinkering with, maintaining, and repairing digital networks and technologies. The connectivity entrepreneurs in the parks were no exception. Pavel, who claimed to be the first connectifier at Parque Trillo, informed me that he used the Internet primarily to watch tutorials and search for information on how to improve his system's configuration for connecting more people through a stable connection. For him, this was a full-time business, earning him approximately 400 CUC per month. However, for others, selling part of their bandwidth was just a side gig. Javiero, for instance, would occasionally visit the park to resell a portion of the thirty hours of Internet he had available through his Nauta Hogar home connection. He explained that sharing half of his allotted hours with other users helped him offset the monthly fee he paid for the service.

Connectify, the software that allowed users to share their laptop's Internet connection with other devices and extend the Wi-Fi range, became an unexpected success story in Cuba and essentially became synonymous with the use of an informal Wi-Fi network. When the U.S. company behind the application learned of its popularity on the island, it even launched a PR campaign around it, announcing that it would provide Cubans with free access to the premium features (knowing, of course, that Cubans wouldn't be able to pay for the service due to the U.S. embargo and would resort to using cracked versions instead).

Connectify was not the sole method used to redistribute the Wi-Fi signal in the parks. Technologically savvy entrepreneurs utilized wireless technologies

such as routers from Latvian company MikroTik, known for its robust Rout-
erOS software system, or access points like Ubiquiti's NanoStation. These tools
allowed them to bridge the park's signal to locations several miles away, offering
connectivity services in neighborhoods that lacked Wi-Fi hotspots nearby. Importing Wi-Fi equipment into Cuba was illegal until 2019. However, there was
a thriving black market for these devices, which were often smuggled into the
country clandestinely by mulas, disassembled or hidden among other goods
such as television sets. Particularly in less affluent neighborhoods like Habana
Vieja and Centro Habana, these informal Connectify networks remained the
primary means of access for many individuals, even after the introduction of
mobile Internet. Maybel, an Afro-Cuban woman in her early twenties working
in the kitchen of a four-star hotel in Centro Habana, highlighted this fact to me
in April 2021:

> Whenever I get the chance, I try to buy mobile data, but now, during the
> pandemic, since the hotel is closed, I can only afford Connectify once a
> week. The guy who sells it lives just around the corner. Sometimes, if I'm
> lucky, the signal reaches my room, but usually, I have to sit on our doorstep
> to get the signal. The connection speed is fine, and for the price of a one-
> hour scratch card, I get a twenty-four-hour connection because he only
> changes the password once a day.

The tarjeta resellers and connectifiers of Parque Trillo and other Wi-Fi
hotspots did not simply sit as parasites on the government-provided infrastructure. Instead, they performed crucial infrastructural work, addressing
its limitations and expanding its coverage. By offering connectivity at more
affordable prices, outside of ETECSA office hours, without the need to wait
in long queues, and in areas not covered by the state provider, they made
Internet access more accessible to the public. Moreover, these individuals
were involved in articulation work, the work "required in order to bring together discontinuous elements—of organizations, of professional practices, of
technologies—into working configurations" (Suchman 1996, 407).[5] This often
overlooked and backgrounded articulation work helped ensure the smooth
operation of the official state infrastructure for ETECSA's users. For example,
when observing tarjeteras like Glenda, a regular presence at the park, I noticed that she spent more time assisting her clients with their phone settings,
clearing caches, and deleting cookies than actually selling scratch cards. Some
users I interviewed, like Paolo, a man in his late fifties, even explicitly expressed their preference for connectifiers over the official service because they

provided support and assistance when encountering difficulties. "If I'm logged in with my ETECSA account and the connection goes down, I lose the time I paid for, but the guys here always let me finish what I wanted to do online and help me if something doesn't work." Tarjeta resellers and Connectify entrepreneurs hence played the role of Internet intermediaries, acting as frontline workers at the interface between clients and the network. Many of my research participants recalled their initial visits to the park when they didn't know how to connect to the Internet. They were quickly directed to these connection brokers who assisted them with turning on their phone's Wi-Fi, finding a less crowded antenna, navigating the unintuitive Nauta login portal, or downloading and setting up essential apps. As unofficial customer service representatives not paid by ETECSA, these intermediaries embodied the Internet for inexperienced users. Their infrastructure work facilitated broader ICT participation by providing new users with the necessary basic knowledge, helping with troubleshooting, and making the Internet infrastructure accessible to a wider public. The interface work performed by these connection brokers, at the boundary between the virtual and physical realms, was a kind of affective labor. They mediated between the technical infrastructure, the materiality of the Wi-Fi signal, and the devices and needs of the users. By positioning themselves between the official Internet infrastructure and the end user, both connectifiers and tarjeta resellers found a way to tap into the corporate state's business model and create a revenue stream for themselves. They created value for their customers by smoothing out the friction that often arose between the user and the unreliable, volatile, and underperforming network.

However, there was a notable economic difference between the two types of Internet intermediaries. Selling tarjetas, requiring fewer technical skills and less investment, was often performed by less affluent individuals, predominantly Afro-descendant Cubans without access to remittances from family members abroad. They were frequently unemployed or underemployed, which allowed them to spend time standing in line outside ETECSA offices. On the other hand, connectifiers needed significant startup capital and expertise to establish their businesses, but their ventures were far more profitable than the resale of top-up cards. The microcosm of connection brokers at Parque Trillo thus reflected larger economic trends in Cuba, where individuals with access to capital had more economic opportunities, while those relying on inadequate government salaries were compelled to take on irregular and informal side jobs, a development that deepened existing structural divides along racial and class lines in Cuban society (Hansing and Hoffmann 2020).

Conclusion

During my subsequent visits in 2020 and 2021, I noticed a significant decline in the number of people at Parque Trillo and other Wi-Fi parks in Havana. This was largely due to the COVID-19 pandemic, as the government discouraged people from gathering in public places. As a result, the government halted its investment in new Wi-Fi hotspots and even closed down some popular locations. The impact of these changes was evident in Parque Trillo. The tarjeta resellers, who used to be a common sight in the park, had all disappeared. With the presence of a newly opened ETECSA sales booth and fewer customers, they had sought out other informal business opportunities. The connectifiers, who previously facilitated Internet access in the parks, were also no longer present. Instead, they shifted their focus to extending the park's Wi-Fi signal to other areas of the city. This shift was made possible by new legislation introduced by the state in the summer of 2019, which legalized the creation of private wireless networks for noncommercial personal use. As long as the connection was not resold, individuals were now allowed to have their own Wi-Fi networks at home or in businesses. Additionally, the importation of routers and networking equipment was also legalized. Hence, while the business model of the connectifiers remained illegal, the new regulations made it easier for them to acquire better and more powerful devices, allowing them to expand their operations to other parts of the city.

As prices for mobile Internet access started to decrease with the introduction of the LTE standard in July 2019, more and more users could afford to be online via their cell phones, making them less dependent on the Wi-Fi hotspots. This constant connectivity led to the growing popularity of applications that relied on uninterrupted Internet access, such as ridesharing services and online delivery platforms. Improved access to the Internet also empowered self-employed individuals to utilize it as a tool for their businesses, contributing to the emergence of a networked public sphere. In chapters 4 and 5, I delve into the intricate social dynamics that unfold from Cubans' engagement with digital platforms. Prior to this, the next two chapters shed light on the ingenious alternative data distribution networks crafted by Cubans. These vernacular infrastructures facilitate the sharing of movies, software, and music and even enable online gaming experiences. Due to the data-intensive nature of these activities, such networks remain vital, maintaining their significance even with the advent of the mobile data service. The prohibitive costs of accessing these entertainment forms via the Internet make these alternative routes not only attractive but also a practical solution for Cubans eager to enjoy digital leisure activities.

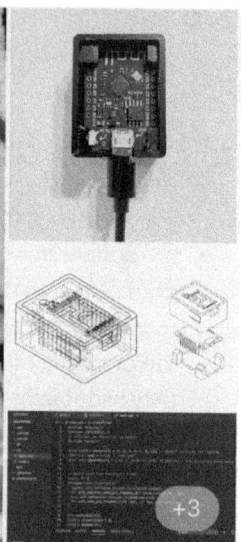

Connectify/Free_Wi-Fi [poesía] (2020–21)

Interactive installation / public intervention consisting
of custom solar-powered Wi-Fi modules
BY NESTOR SIRÉ

www.da.gd/wi-fi

Participating poets: Ismaray Pozo, Mario Espinosa, Gabriel Ojeda-Sagué,
Oscar Cruz, Jamila Medina Ríos, Legna Rodríguez Iglesias, Lisann Ramos,
Javier L. Mora, Adonis Ferro, Lizabel Mónica, Katherine Bisquet, Kyle
Carrero Lopez, Sindy Rivery Elejalde, Martha Luisa Hernández Cadenas,
Cuci Amador, Yosie Crespo, Julián Bravo Rodríguez, Ricardo Mayo,
Randy Amor, and Ylena Zamora-Vargas.

Concept codeveloper and software: Luis Rodil-Fernández
Project assistance: Yainet Rodríguez
Copyediting: César Segovia
Graphic design: Noah Levy
Hardware development: Copincha
Hardware design: Maurice Haedo Sanabria
Hardware electronics: Lázaro Alejandro Navarro Méndez
3D printing: Eduardo Pujol

FIGURE 1.2. Installation views of *Connectify/Free_Wi-Fi [poesía]* from various locations in
Little Havana, Miami, and Centro Habana, Cuba.

This public intervention by Nestor Siré, commissioned by the poetry festival O-Miami and primarily developed during a Fountainhead artist residency, explores the effects of public Wi-Fi access on the relationships between Cubans residing on the island and those who have emigrated. I had the privilege of following the creation and development of the project up close and helped by bringing some of the necessary technical components to Cuba. The project engages with a common occurrence at Cuban Wi-Fi parks, where numerous networks appear in the available network list on devices, each with a name different from the uniform WI-FI_ETECSA provided by the national monopolist ETECSA. These networks, often bearing humorous names, are the hotspots established by the informal connection brokers operating in the parks who redistribute ETECSA's Wi-Fi signal. Nestor's intervention involved using the SSIDs (service set identifiers—the sequence of characters that uniquely names a Wi-Fi network) of such private networks to showcase lines of poetry. These poetic lines became visible whenever individuals accessed the list of available hotspots on their mobile devices.

To accomplish this, Nestor installed modified Wi-Fi modules at public spaces in Havana and Miami. These devices were codeveloped by the members of the Havana-based hacker and maker collective Copincha. They were specifically designed for outdoor use and were equipped with a range of features to ensure durability and functionality. These include a security system, a waterproof 3D-printed box, and a self-sustainable renewable power system. The power system utilizes a solar panel and repurposes the charging port of a recycled power bank, allowing the device to operate autonomously for over five months without human intervention. The software installed on these devices is a Wi-Fi spoofing tool, typically associated with hacking and state espionage activities. However, in this project, the Wi-Fi spoofing software simulates the presence of legitimate Wi-Fi networks not to steal people's data but rather to infuse the radio waves with poetry.

The poems themselves were contributed by twenty young Cuban poets who represent the same generation but are scattered across various locations. Some of them reside in Cuba, while others live in the United States, Latin America, Spain, or China. Additionally, among the contributing poets are individuals who are first-generation Cuban Americans born in Miami. The poems incorporated into the project were carefully crafted to meet two specific requirements. First, they were limited to a maximum of thirty characters, mirroring the maximum possible length for a custom SSID. Second, the poems aimed to convey the individual perspectives and experiences of their authors regarding Cuba, capturing their nuanced emotions, reflections, and aspirations related to the country.

By replacing traditional Wi-Fi networks with these bilingual micropoems, the project seeks to engage in transnational communication and bridge the separation that exists between Cubans on the island and those who live in the diaspora. Just as the introduction of Wi-Fi hotspots in public parks provided an opportunity for many Cubans to reconnect with their family members who had left the country, this poetic intervention serves as a means to facilitate dialogue and connection across political and ideological divides. Literally waiting to be transmitted as a signal, these poetic expressions of Cuban cultural identity reached thousands of users and potential readers. By utilizing poetry as a medium, the project aimed to transcend the limitations of the toxic and heavily polarized political and media narratives on both sides of the Straits of Florida, providing a space where individuals could explore and share their personal perspectives, emotions, and experiences related to Cuba.

The QR code provides access to a comprehensive data package featuring exhibition views, the open-source software driving the installation, a detailed construction manual for the solar-powered Wi-Fi module, the exhibition brochure, and video documentation.

CHAPTER 2

Hard Drives

On a balmy February evening, right after a *frente frío* (cold front) had released Havana from its grip, Nestor and I embarked on our way to our friend Chuli's house, eagerly anticipating a special communal viewing experience. With the responsibility of fetching pizzas from our favorite Italian eatery falling to me, Nestor had ventured to a neighborhood copy store to fill a flash drive with fresh content. It was the evening before the Oscars, and as we walked down the potholed concrete runway of Calle Infanta toward Parque La Normal, Nestor animatedly shared his comprehensive assessments of each film in the "best picture" category. I could only listen attentively to his exhilarating lecture as most of the films had yet to hit the screens of German cinemas. Nestor, on the other hand, had already devoured almost all the nominated films and had a clear idea of which ones were most likely to satisfy what he thought were the rather predictable tastes of the Academy members. My curiosity was piqued, and I inquired if we would be watching any of the nominated films that evening. Nestor's grin grew wide as he responded. Instead of indulging in cinematic achievements, we were going to enjoy a new guilty pleasure shared by him, Chuli, and their partners: *La Isla de las Tentaciones*. This trashy Spanish reality show had become a cultural phenomenon among young Cubans. Its premise was simple yet intriguing—couples at a crossroads would embark on a tempestuous journey of self-discovery, flirting with the single life in an attempt to determine the fate of their relationships. It had gained unprecedented popularity, captivating the imaginations of Havana's youth with its mix of drama and emotional turmoil. Despite the latest episode having aired

FIGURE 2.1. Informal media distribution in Cuba has evolved from renting cheap novels to sharing VHS tapes and self-burned DVDs, culminating in the island-wide grassroots offline data distribution network, el paquete semanal. This one-terabyte data collection is meticulously organized into a consistent structure of files and folders and is distributed through neighborhood copy stores and a network of delivery persons.

just the previous night on Spanish television, Nestor had already managed to secure it on his trusty USB stick.

In spite of the restrictions and limitations on Internet access in Cuba, Cubans have found a way to have instant access to the latest international movies, TV series, YouTube videos, software, and music albums. While government-provided public Wi-Fi and mobile Internet are not suitable for data-intensive activities like streaming video or downloading large files due to high costs and often slow bandwidths, Cubans are still able to engage with current global media content. This is made possible through an ingenuous alternative data distribution network that relies on (mostly pirated) digital materials that are centrally downloaded by individuals with privileged access to the official government Internet infrastructure and then made available to the entire population. At the heart of this vast informal business operation are compilation studios known as *casas matrices*. These studios organize the daily download of hundreds of gigabytes of data by mobilizing a network of people working at government workplaces, universities, and tourist hotels. The resulting data files are then circulated nationwide as a curated, one-terabyte content collection known as el paquete semanal. A sophisticated human infrastructure of delivery persons, so-called *paqueteros*, ensures that this content reaches even the most remote corners of the island, delivering it directly to customers' homes on hard drives. Furthermore, individual media files, such as the latest episode of a popular series, are sold in puntos de copia found in every neighborhood of virtually every city.

Once a customer has paid to copy the data, these files often continue to circulate within networks of friends and families, transferred via USB sticks or through peer-to-peer file-sharing applications like Zapya. This ensures that every Cuban can stay up to date with international entertainment. Interestingly, local content producers have also begun to leverage these offline distribution structures to circulate their own music, videos, and journalism throughout the island, offering an alternative publishing platform outside of the government-controlled media infrastructure. El paquete not only compensates for the limitations of Internet access but also addresses the absence of entertainment and the exclusion of independent content on Cuban television. Since the nationalization of all mass media after the Revolution, Cuban television has served as the central government-controlled tool for education, culture, and political mobilization, promoting socialist ideals (Concepción Llanes and Oller Alonso 2019). As a result, el paquete enables Cubans to reduce their reliance on the official mass media, which, according to the Cuban constitution's prohibition of private media, remains concentrated in state hands.

The Emergence of Alternative Media Distribution in Cuba

Vernacular networks for media distribution, offering alternatives to the state-run mass media's propaganda and limited access to foreign entertainment, have existed in Cuba for several decades. These networks serve as informal means of reconnecting the Cuban people to global media sources from which they have been politically and economically disconnected. An example of such a distribution network can be traced back to the 1970s when Evelio, the grandfather of my frequent collaborator Nestor Siré, operated a successful business in his hometown of Nuevitas in Camagüey province, renting out foreign paperbacks that were in high demand. These paperbacks, which included popular Western and romance novels by authors like Donald Curtis and Marcial Antonio Lafuente Estefanía, were often printed in Mexico and unavailable in Cuban stores. Evelio's business thrived through his resourcefulness in acquiring new material for his clients. He would occasionally travel to places like Sancti Spíritus, Ciego de Ávila, or even Havana to trade novels with other collectors, ensuring a fresh supply for his customers. This example demonstrates how Cubans have always managed to find ways of accessing foreign media, even during periods when the government prohibited the likes of the Beatles due to concerns of *diversionismo ideológico* (ideological diversion) or decadent American influence.

By the 1990s, Evelio and his son Ever had adapted their business to the evolving technological landscape. Ever, explaining their transformation, stated,

> Cuba was on the verge of a technological revolution. VCRs, both Beta and VHS equipment, were imported from various countries, particularly from Mexico. During that time, we still had black-and-white TVs, specifically the Krim-218 model, but VCRs were already making their way into the country. It was an incredibly swift leap forward. And then the so-called *la chopin* stores opened in our area, government stores where you could purchase imported products in dollars. This marked the beginning of a movement towards paying in foreign currency. They started paying huge salaries to the workers in the labor centers, especially to the workers in the thermoelectric plants, because they worked with Cuban crude oil. Workers who previously earned 200 to 300 pesos now found themselves with an additional hundred dollars or more to spend at these stores. So, people began buying color TVs from the shops, and Betas and VHS through the black market. This transition happened rapidly, as people essentially stopped renting novels and instead started watching movies and telenovelas on cassette tapes. We realized that it was important to grow with cassettes, and we started selling our novels and buying cassettes.

As the popularity of VCRs grew, an informal video rental system emerged in Cuba. While blank cassettes could be purchased at *la chopin* (Cuban slang derived from the English word shopping), acquiring foreign media content for them required alternative methods. Initially, Evelio and Ever leveraged their connections with employees at tourist hotels, particularly technicians who had access to U.S. and Latin American television stations via the hotels' satellite antennas and would record movies and TV series for them. However, a clandestine market for illegal satellite television reception soon surfaced. People began discreetly installing antennas on their rooftops, often concealed within water tanks, and then offering access to satellite channels through coaxial cables clandestinely laid throughout their neighborhoods. For a monthly rental fee, individuals could access this illicit satellite TV service. Some of the owners of these rooftop antennas also started recording movies and TV shows, initially using VHS recorders and later transitioning to computers with video capture cards. They would then sell these recorded materials to video rental entrepreneurs like Evelio and Ever. Some of the earliest casas matrices began curating thematic collections of movies and series on cassettes. They even developed recognizable brands by adding animated logos at the beginning of their recordings.

The video rental business carried greater risks compared to book rentals, as the police actively targeted private possession of antennas. During the Special Period in the 1990s, the Cuban government also intensified its crackdown on black market activities and illegal businesses through its Operación Maceta policy.[1] Despite these dangers, many Cubans were willing to take the risk. In a time of economic hardship and political repression, engaging in such infrapolitical acts of economic resistance offered the potential for substantial incomes.

To minimize personal risk and avoid police scrutiny, Evelio and Ever took precautions when copying tapes for distribution. They carefully removed the advertisements that interrupted the foreign films and TV programs. These commercials portrayed the economic reality outside of Cuba, and they were aware that the state did not want its citizens exposed to capitalist products and market prices. The individuals involved in importing VCRs or recording and copying foreign television programs were often highly skilled or educated individuals with opportunities to travel and make contacts outside of Cuba. They included touring musicians, pilots, and doctors participating in international missions to countries like Venezuela or Brazil. These individuals would invest their hard-earned foreign currency in electronic devices that they brought back to Cuba to supplement their meager government salaries (Andaya 2009). Doctors and other highly skilled workers, given their social standing, faced less harassment from the police when engaging in these activities. As Anna Pertierra (2009) observed in her study on informal VCR distribution and consumption circuits in Cuba, although importing VCRs or operating video stores was illegal from the government's perspective, it was not seen as socially or morally wrong by consumers. Many of her research participants, who were also involved in community revolutionary activism, were regular customers of these video rental businesses and avid consumers of capitalist media products like Mexican telenovelas and Hollywood films. Video rentals thus provided a form of leisure that, while formally illegal, offered a respectable retreat into the domestic space.

By the mid-2000s, there was a notable relaxation in the political situation in Cuba, leading to the legalization of importing electronic devices by the government. This change allowed individuals to officially bring mobile phones, computers, and DVD players into the country. With this new development, the video banks swiftly adapted their operations from copying VHS tapes to burning DVDs that were more cost-effective to purchase and easier to transport. In the late 2000s, as the Cuban government introduced labor market reforms and issued cuentapropista licenses for self-employed individuals, video bank owners could finally legalize their businesses by applying for a *comprador-vendedor de discos* (buyer and seller of discs) license.[2] This shift enabled video banks, previously operating clandestinely from private residences or solely as delivery services, to open regular stores where discs could now also be sold instead of rented.

As the number of computer owners increased and the prices of storage media like hard drives and USB sticks decreased, data copying became more widespread and accessible. Casas matrices, the central hubs responsible for compiling content, started receiving their content from piracy websites. IT students at technical universities and administrators at large state-owned enterprises with Internet access acquired media content through file-sharing networks, using their individual employee Internet quotas to facilitate these downloads (Pertierra 2012). With the rise of computers and TVs equipped with USB connections that allowed users to watch movies as video files, traditional physical media such as cassettes and discs became obsolete. Matrices recognized the growing demand and profitability of broadening their customer base. Instead of exclusively targeting small business owners who burned DVDs for individual customers, they realized that it would be more advantageous to sell their downloaded files to a wider audience simultaneously. This marked the beginning of the concept of el paquete semanal. Rather than purchasing individual movies or TV series on discs, consumers now had the option to buy a comprehensive collection of data organized by topic and divided into different folders, encompassing the entire content that the matrices had downloaded. Initially, these collections were compiled on a monthly basis, but as Internet access improved and more hard drives became available, a weekly release schedule became the standard.

El Paquete's Content

The majority of el paquete's weekly content, totaling one terabyte, consists of a wide range of movies, TV shows, music, and various forms of Internet content. While the collection includes a significant number of Hollywood films and Netflix and HBO productions, it also offers a diverse selection of content from different parts of the world. This includes Latin American, Spanish, and Turkish telenovelas, Japanese anime, and Korean dramas. The music section primarily features local and international reggaeton, a genre popular in Cuba but officially banned by the state mass media due to perceived moral concerns (Levine 2021). A considerable portion of the material is sourced from YouTube, encompassing popular video blogs on fitness, technology, and gaming, makeup tutorials, and comedy shows. In addition to audiovisual content, el paquete incorporates various software such as mobile apps for iOS and Android, pirated video games, antivirus programs, PDFs of international magazines and newspapers, and offline archives of local classified platforms like the Cuban version of Craigslist called Revolico (hosted in Spain and often blocked in Cuba). The selection of online content included in the paquete has evolved over time to adapt to changing Internet access in Cuba. As shown in Julia Weist and Nestor Siré's (2020) long-term quantitative study of the transformation of el paquete's content, the collection shifted to reflect the increased availability of social media. For example, it featured fewer of the formerly ubiquitous screenshots of posts from prominent figures such as actors or musicians once mobile Internet became more accessible.

The online materials downloaded by the matrices for their customers are organized into alphabetically ordered folders, often with numerous subfolders, which may vary in number and naming depending on the paquete version. Popular genres like Korean dramas (referred to as *doramas*), telenovelas, and reality shows typically have dedicated folders. Sometimes, further categorizations are made based on whether a series is a "premiere," "in broadcast," "finished," "classic," or "dubbed in Spanish." Within each series folder, one to five consecutive episodes are typically included, ensuring that customers are entertained until the next paquete arrives. Films are organized into folders labeled as "premieres," "in HD quality," or "documentaries." Matrices also compile curated folders, showcasing films of specific genres or retrospectives of particular actors. The *humor* and *interesantes* folders contain Internet content like adorable pet videos, pranks, or a selection of popular Latin American YouTubers. The *deporte* (sports) folder features material from major sporting events such as baseball, boxing, and European soccer matches.

The individuals working for the matrices as downloaders and compilers have specialized roles in handling specific types of content. They are responsible for identifying and downloading new materials, naming the files, and organizing them into the relevant folders. The selection of content for el paquete is based on past popularity, global trends, and occasionally consumer requests communicated through email or paqueteros. The paquete's content is highly current, with popular television shows and movies often available for distribution as early as the day after their release or airing.

El paquete, which originated from informal media distribution networks enabling access to foreign entertainment, has now become a local content distribution platform for Cuban creators such as musicians, YouTubers, and independent journalists. Matrices typically offer a subscription model for hosting content. The YouTubers whose trajectories I follow in chapter 5 have long paid 2 CUC per video or a flat rate of 10 CUC per month to circulate their videos offline on the island. To diversify the content, some matrices include initiatives like Nestor Siré's *sección arte* (art section), a monthly folder serving as a digital exhibition space featuring art-related documentaries and open calls for festivals or residencies, or an activist-curated LGBTIQ+ section for free. Independent Cuban PDF magazines, including *Vistar* (a pop culture magazine) and *Negolution* (a business magazine aimed at self-employed individuals), are also disseminated through the network. However, to safeguard their business, the matrices do not accept political or antigovernment material, thus not providing a platform for independent online journalism projects like *Periodismo del Barrio*, *La Joven Cuba*, or *El Estornudo*, which aim to offer alternative perspectives on contemporary Cuban reality. These projects' websites are usually also geoblocked in Cuba (Henken 2021a).

Some of the curators responsible for individual folders within the paquete have gained recognition as gatekeepers, providing an avenue for unknown content producers to gain national exposure. For instance, Abdel la Esencia, a well-known local producer, curates the popular *música nacional* section for the matriz Odisea. He includes his contact information, name, and image in the files' metadata and as a watermark in the videos, allowing aspiring artists to reach out to him for promotion within the paquete. In addition to his curatorial role, Abdel also works as a promoter for individual musicians and events, and he even sells advertising space within his folder. Given that reggaeton is not played in state media and the paquete serves as the primary distribution medium for this genre, aspiring musicians heavily rely on Abdel to advance their careers.

Publishing local advertisements serves as an important additional source of income for the matrices. Omega, one of Havana's two leading matrices, operates its own advertising agency, offering advertising space within the paquete. Since the media cannot be used to promote capitalism under the constitution, advertising is technically banned in Cuba. Consequently, for a long time the paquete remained one of the few platforms where Cuban small business owners could promote their services, making it a key element of the country's reluctant economic liberalization and an important economic space. Advertisements are integrated into the paquete in the form of JPEG files or as commercials inserted at the beginning or end of popular TV series episodes or as standalone MP4 files dispersed throughout various folders. Surprisingly, a significant number of Miami-based Cuban American businesses, such as Ñooo Que Barato and Valsan, also place ads. These stores specialize in inexpensive, Chinese-made clothing or electronic devices frequently sent as remittances from the diaspora to relatives on the island. Through the paquete, these businesses can directly target recipients of such material remittances. For example, Leyian, a cousin of a Cuban friend who had emigrated to Miami six years ago, recalled an incident where his aunt in Havana, who had never been to the United States, asked him to purchase clothes for her from La Cuevita de San Miguel, a store in Hialeah that he had never heard of.

However, the significance of the paquete as an advertising platform has recently somewhat diminished. With the widespread availability of mobile Internet, many Cuban entrepreneurs now prefer to engage in online marketing and collaborate with local social media influencers who have large followings on platforms like Instagram, TikTok, and YouTube (see chapter 5). Paid endorsements, product placements, and testimonial advertising through these channels have led to reduced investment in paquete advertisements, various copy store owners confirmed to me.

El Paquete's Distribution System

Two matrices, Omega and Estudios Odisea, both based in Havana, dominate the market and sit atop el paquete's distribution pyramid. In the past, they were the sole entities responsible for centrally organizing the download and distribution of data throughout the country. While the type of content they offer is similar, Omega is renowned for providing higher quality movies and series, whereas Odisea offers a wider variety of MP3s and music videos. Both matrices have shifted away from compiling weekly content collections and now sell daily updated content to sub-distributors, referred to as sub-matrices, located throughout the island. These sub-matrices combine the data received from both Omega and Odisea to create their own individual paquete semanal, which they subsequently sell to neighborhood paqueteros responsible for delivering the final product to end customers. Often, sub-matrices organize their paquete according to a unique local structure, utilizing their own name and brand. Additionally, the two main matrices directly target neighborhood puntos de copia, which are selling individual files to customers who prefer not to wait for the next paquete release. These customers may be seeking the latest episodes of their favorite series or older seasons of shows previously included in the paquete weeks ago. Sub-matrices and puntos de copia often operate as the same business, with sub-matrices assembling weekly paquetes for self-employed paqueteros and simultaneously operating copy stores for walk-in customers purchasing individual files. Despite the shift toward daily updates to cater to the demand for freshly published content, the format of the weekly curated data collection, totaling a terabyte, remains unchanged. Although the matrices now offer a significantly larger volume of data, the size of the paquete itself has not increased. The limiting factor in this vernacular infrastructure is the size of the hard drives owned by paqueteros, which typically do not exceed one terabyte. Given that such devices represent a substantial investment in Cuba, it is unlikely to see a change in the near future.[3]

The paqueteros serve as the foot soldiers in el paquete's distribution network. Typically, they own a few hard drives onto which they copy the paquete purchased from a sub-matriz and then deliver it to their customers' homes for a fee of the equivalent of two to three dollars (2 CUC before monetary reunification, or 75 CUP before rampant inflation). After twenty-four hours, when the customers have copied the desired content for the week, the paqueteros retrieve the hard drives and pass them on to the next customer. Similar to the tarjeta vendors in the Wi-Fi parks mentioned in the previous chapter, most of the paqueteros I encountered did not have academic education, were often of Afro-Cuban descent, and relied on this informal work either as supplementary income to their government salaries or as their primary source of livelihood. The paqueteros I followed typically served between ten and thirty customers per week and owned between three and ten hard drives.

Like the tarjeteros and connection brokers in the parks, paqueteros also performed essential interface work for their customers. They didn't just hand over a hard drive; they often personalized the paquete according to their customers' individual tastes and technical capabilities. For instance, Albany, a part-time musician who delivered the paquete, had predominantly older subscribers with whom he would sit down and recommend specific content to prevent them from feeling overwhelmed by the abundance of offerings. Miguel, a paquetero from Habana Vieja, had customers who did not own a computer. In such cases, he prepared a selection of films and series on a USB stick that could be directly plugged into their television. Other paqueteros did the same for restaurant owners who had installed a TV screen to entertain their patrons with the latest music videos, saving them the hassle of transferring the files to a USB stick. If customers were searching for a particular movie or show not available in the collection, or if they were behind on a popular series and needed access to earlier seasons, many paqueteros would go the extra mile to obtain those materials, either from their sub-matriz or from renowned copy stores known for their extensive archives.

Most paqueteros cater to a specific neighborhood, and newcomers to the business are more inclined to attract new customers rather than compete directly with established paqueteros in the same area. Being neighborhood-based entities, they often develop close relationships with their customers. Similar to Netflix's recommendation algorithms, paqueteros learn their customers' preferences over time, allowing them to provide more suitable content recommendations. Thus, small additional tasks like finding a specific movie or preparing data tailored to the customer's technological setup are considered investments in building and maintaining customer relationships.

The nationwide distribution of the data collection, which guarantees reliable access to the latest content even for paqueteros in the most remote villages, is coordinated by the two matrices based in Havana. Each morning, a fleet of messengers embarks on a mission to transport the hard drives containing the daily updates to all major Cuban cities using the national intercity bus network. From these regional hubs, the data are further transported to the farthest corners of the island using government or private taxis as well as local buses. In many cases, bus drivers themselves seize the opportunity to earn extra income by making unscheduled stops at intersections where someone in a car or on a motorcycle awaits to collect the drives and deliver them to villages that are not connected to the public transportation network.

The value and price of the paquete depend on the freshness of its content, as more recent offerings are more enticing to potential customers. Time plays a crucial role in this business, as many end customers eagerly await the latest episode of a popular TV show or series or an important sporting event. Consequently, the closer the paquete is to Havana in the distribution chain, the higher its price, which gradually decreases with each passing day and each intermediary involved in its transfer. The two major matrices charge the equivalent of ten dollars for their daily update and cater to hundreds of customers across the island. However, due to strict Cuban regulations aimed at preventing private companies from becoming too large, they are unable to establish branches in other cities, rendering centralized business operations or nationwide monopolies impossible. As a result, it is plausible that certain local sub-distributors purchasing from the main matrices generate more profit by selling data in their respective regions compared to the matrices based in Havana. Nonetheless, it is this highly decentralized and personalized distribution model that contributes to el paquete's extensive reach throughout the country.

As the paquete travels through the hands of messengers, sub-matrices, and paqueteros, the content undergoes modifications to cater to different local or individual preferences. Regional sub-matrices mix the updates from the main matrices, occasionally supplementing them with their own downloaded material or specific local content. They often remove Havana-specific advertising and replace it with ads for local businesses. Consequently, there exist countless variations of the paquete. Thus, it is not accurate to refer to el paquete in the singular form, but I do so because that is how Cubans commonly refer to it.

In nearly every town and every neighborhood of larger cities, one can find puntos de copia where individual media files or the entire paquete can be purchased or where cell phone apps are installed for customers. These small businesses often operate in the owners' living rooms or hallways, transforming their private spaces into quasi-public areas where they receive their customers. For instance, I frequently visited Cinemania 4K, a store near Parque Trillo in Havana. It began as a venture run by Osvaldo, the owner, with only one computer placed near his apartment window, where long lines of customers would form to copy new films. Over the years, the business has undergone remarkable growth, and it now occupies a fully renovated, air-conditioned storefront with six employees accessing 200 TB of data to fulfill every conceivable need of their customers. Osvaldo gradually expanded his extensive archive by purchasing large data collections from other copy stores in different parts of the city, avoiding direct competition. With his business acumen, Osvaldo eventually ceased selling the entire paquete because he deemed it too inexpensive (he could earn as much selling five individual films as customers paid for the entire data collection) and because copying a terabyte of data would tie up one of his USB hubs for at least three hours.

To enhance their business administration and facilitate large-scale data copying, puntos de copia and sub-matrices rely on powerful Cuban-developed software solutions. One such program is *Paquetecopies*, which enables simultaneous copying from a single source to multiple destinations at maximum speed. During the copying process, devices can be automatically synchronized, ensuring that additions or deletions of files and folders are replicated on all target devices. Another software called *MiRON* monitors the copying process, enabling copy store operators to maintain records of the number of customers served, files copied, and devices used. It also automatically generates receipts for clients, listing the copied files, and stores them on the clients' devices.

Despite not directly purchasing the paquete or visiting copy stores, many individuals still receive its contents. Groups of friends or colleagues often pool their resources to share the cost of the paquete. Additionally, the data collection is disseminated through FTP (File Transfer Protocol) servers on neighborhood computer networks, as discussed in chapter 3. Offline file sharing is also widespread in Cuba, with paquete consumers sharing its contents with their neighbors, friends, or schoolmates. Many Cubans always carry USB flash drives with them, enabling them to exchange data with friends or acquaintances whenever the opportunity arises. At events like the science fiction and fantasy convention I attended at the Centro Hispanoamericano de Culturas in Havana, organizers set up computers with extensive collections of movies, anime, and e-books, allowing attendees to copy as much content as they desired.

Another popular method of sharing data is through the mobile app Zapya, which facilitates file transfers between smartphones without the need for cables, Wi-Fi, or mobile data. Zapya leverages the phone's Wi-Fi tethering or hotspot feature. Prior to the availability of mobile Internet, Cuban youth utilized Zapya not only for sharing songs, pictures, and short videos but also as a local substitute for messenger apps like WhatsApp (although users must be physically near each other). These examples illustrate that in Cuba digital information is frequently transmitted through face-to-face encounters rather than relying solely on networked technologies. Movies, series, and music files circulate within family and friendship networks via USB sticks and phones, which people use as storage devices. Therefore, people's interactions with digital media almost always incorporate a social element.

El Paquete as a Human Infrastructure

El paquete operates as a large-scale business operation, providing income for thousands of individuals involved in its distribution network, including paqueteros, downloaders, copy store owners, and their employees. However, it is also a people-embedded infrastructure, where countless individuals collaborate to ensure that millions of Cubans have access to global and local media content that would otherwise be unavailable to them. As I mentioned earlier, participants in this informal distribution network often engage in additional unpaid or backgrounded articulation work that ensures el paquete reaches all layers of Cuban society (Dye et al. 2018). Paqueteros, at the interface between the network and the customers, not only deliver the content but also provide recommendations and even preselect materials based on their clients' preferences. They make an extra effort by seeking additional requested content and sometimes enabling access for those who don't own computers. Cuban programmers develop software to enhance the stability and efficiency of the copying process. The matrices in Havana coordinate a network of messengers and intercity bus drivers to distribute el paquete across the entire country, even though some regional distributors may earn more money from their data. And end users actively share the content with friends and relatives who cannot afford to purchase the paquete, ensuring that no Cuban is left without access to its content.

Nevertheless, the network's heavy reliance on human actors, their relationships, and their cooperation as well as the limitations of their technical equipment also expose certain vulnerabilities. Due to the significant financial stakes involved, rivalries often emerge between different sub-matrices and copy stores competing for market share. For instance, the owner of Omega, who used to work for the Odisea matriz, decided to establish his own business, leading to ongoing conflicts over customers, advertisers, and content providers. This rivalry once escalated to a physical confrontation, prompting both matrices to publish their perspectives on the conflict in their paquete editions. Furthermore, there have been instances of copy stores attempting to pay the matrices to exclude local competitors from the distribution network. However, due to the ease of data copying and the decentralized nature of the distribution chain, permanently excluding new vendors seeking to enter the business is almost impossible.

Moreover, el paquete is characterized by its ephemeral nature. A significant portion of its content is lost within a short period of time. This is primarily due to the limited storage capacity of the paqueteros, who typically have terabyte-sized hard drives, and the puntos de copia, which cannot retain all the data they receive. Osvaldo, the owner of Cinemania 4K, is an exception as he stores the daily updates from the matrices for a few days. However, even he eventually deletes certain content to make room for new additions. He starts by removing folders containing YouTube content, followed by sports and most music, and retains only movies and series for a longer duration. Consequently, without coordinated data archiving and backup servers, a significant portion of the materials downloaded by the matrices disappears within weeks or months.

The unreliability of el paquete's content, which can potentially change as it passes through different intermediaries, poses a constant challenge for local producers aiming to reach a wider audience through the distribution network. For example, Ernesto, a twenty-year-old YouTuber from Alamar, located on the eastern outskirts of Havana, decided to stop distributing his videos through el paquete. His content rarely reached his neighborhood as local paqueteros consistently prioritized other materials they deemed more relevant for their local clients. Similarly, members of Luz Visión, a production company specializing in evangelical audiovisual content, expressed frustration over the sometimes sporadic distribution of their content. They found that their folders were frequently deleted or overlooked by paqueteros or sub-matrices. At times, intermediaries intentionally hinder the circulation of specific materials. A few years ago, the curator of a folder containing cell phone applications attempted to introduce a new business model within the paquete infrastructure. He paid both main matrices in Havana to include his folder but protected it with a password. Customers who wanted access to the contents received the password via SMS after transferring him a CUC as phone credit. However, various sub-matrices promptly undermined his efforts by circulating the password in a .txt file or deleting the folder altogether.

Various key actors within the human infrastructure of el paquete have the power to either facilitate or impede the flow of information within the network. The main matrices, concerned about potential government repression, choose not to distribute most of Cuba's independent digital journalism. On the other hand, influential curators like Abdel la Esencia can elevate the careers of specific reggaeton musicians by promoting their work in their curated folders. Additionally, a particular sub-matriz, acting as the sole top-level distributor in an area, has the ability to effectively block specific content from reaching an entire city or province if they decide not to pass it on.

The fragility of el paquete's infrastructure became evident during the disruption caused by the coronavirus lockdowns, which severely limited public transportation between provinces, the backbone of its offline national distribution system. In response, distributors had to find alternative methods to transport the data. For instance, they resorted to utilizing Medibus, a bus primarily designated for transporting individuals requiring hospital care and hospital staff from remote locations. Despite these efforts, many puntos de copia and paqueteros in the provinces faced challenges in maintaining the same level of service as before the pandemic. Yomil, the owner of a copy store in Ciego de Ávila, who purchased the paquete from a local sub-matriz connected to Havana, provided insights into this situation:

> It was disastrous here. Numerous folders were missing, and there were missing episodes of popular telenovelas. While the daily updates were still somewhat maintained, the quality suffered greatly. I had a customer who asked for the twenty-eighth episode of a series, and I had to inform him, "You see, the twenty-ninth episode has been released, but there were issues with downloading the twenty-eighth. Please come and pick it up on Saturday when I receive the next delivery."

These ongoing issues prompted certain local sub-matrices to recognize the necessity of reducing their dependence on the main matrices in Havana. Even prior to the pandemic, sub-matrices in locations such as Bayamo, Camagüey, and Santiago de Cuba had established their own infrastructure for downloading content, taking advantage of improving Internet accessibility and declining costs. Rafael, aged forty-four, the proprietor of Cubaneo Estudios, a sub-matriz in Camagüey that also operates a copy store, identified the primary motivation for undertaking independent content downloading: the desire to acquire the latest content as swiftly as possible.

Havana produces exceptional content, but here's the issue: customers demand fresh material, and they want it quickly. Let's consider the example of *La Casa de Papel*. When a new episode airs, the customer, in this case, the punto de copia responsible for selling it to the end customer, desires it the next the morning. However, the data from Havana, which needs to be transported via bus, train, or plane, takes a significant amount of time to arrive. That's where these numerous small matrices across Cuba come into play, sourcing that content for you. These smaller matrices have experienced growth in terms of download volume, video capture volume, and more. We're not competing with Havana, but we're also generating compelling content. In many provinces, distributors acquire content from Havana to assemble the paquete, but they still seek the exclusive content, ranging from 100 to 200 gigabytes, procured by us smaller matrices. This exclusive content is highly sought-after, encompassing sports, series, soap operas, contests, and more.

As I observed Rafael's daily routine in his home office, where he downloaded, sorted, and copied data, I was amazed by the meticulous efficiency of his workflow, which he had honed during the lockdown. Despite his background as a trained doctor, Rafael considered himself a computer geek and had taken significant steps toward achieving independence, not only from the matrices in Havana but also from his coworkers. Working simultaneously on two computers for often fifteen to sixteen hours a day, he had devised a semi-automated work process aided by various software, relieving him of tedious tasks such as transcoding videos into a consistent format, creating folders, and translating subtitle files. He frequently expressed dissatisfaction with the disorganized folder structure of the Havana matrices and liked to compare himself to a master chef who keeps some secret techniques to himself rather than sharing them with his subordinates.

Rafael had discovered a method to merge multiple Nauta Hogar accounts, pooling their bandwidth to achieve download speeds that were a dream for most other Cubans. This enabled him to download 150 to 200 GB of new material each day. He avoided peer-to-peer platforms that used the BitTorrent protocol and instead used a credit card issued to his father in the United States to pay for premium accounts on file hosting services known for hosting large amounts of pirated content. This approach allowed him to achieve better download speeds. Additionally, he invested in a virtual server in the United States, granting him access to websites that were blocked in Cuba due to the U.S. embargo. Although his monthly expenses were significant, they proved worthwhile. Throughout the pandemic, Rafael had established himself as a distributor of a weekly paquete and daily updates, serving customers in remote locations such as Sancti Spíritus, Santa Cruz del Sur, and Las Tunas. In addition to running a bustling copy store set up in an annex to his house, he also engaged in a side business of downloading pirated video games for stores that sold PC and console games. The limitations on movement and travel imposed by the pandemic had provided him with the opportunity to become an essential node in a reconfigured distribution infrastructure, where Havana had relinquished some of its dominance.

El Paquete's Social Significance

As I have previously argued (Köhn 2019), el paquete offers Cuban consumers a rare experience in a context marked by persistent scarcity. During my interviews with paquete users, a recurring theme that emerged was the sense of abundance described by many participants. They expressed great enjoyment in the vast array of content available through the paquete, which provided them with seemingly limitless choices. In contrast to the limited entertainment offerings of state television and previous informal distribution networks, el paquete offered more content than one could possibly consume in a week.

By liberating viewers from the strict programming schedules and unappealing content of state TV stations, el paquete allows users to watch their favorite TV shows whenever and for however long they desire. Additionally, el paquete has played a significant role in the emergence of various subcultures, such as otakus, cosplayers, and K-pop fans, who rely on it as their sole access to Japanese and Korean pop culture (Humphreys 2021). Cuban youth who read or hear about a new anime series or Netflix show may lack access to global streaming platforms or the means to download it from torrent sites due to technical limitations or the cost of data packages. However, they can request such content from their local copy store or paquetero, who will make it available at a reasonable price. Thus, el paquete creates a space where individuality can be expressed through consumption, enabling consumer citizens to develop a global perspective and choose from a wide range of international media content. This experience of choice and abundance sharply contrasts with everyday life in Cuba, which is characterized by scarcity of consumer goods and the need to wait in long lines, often for hours, to purchase even the most basic food items. This reality restricts individual consumer choice due to the limited options available.

The significance of el paquete as a social lifeline became particularly evident to me during the COVID-19 crisis. As strict lockdown measures confined most of my research participants to their homes, el paquete emerged as a source of solace, combating boredom and providing an escape from the challenging economic conditions in the country. In the words of one interviewee, it was instrumental in helping her survive the pandemic. While much of the world shifted social interactions to the digital realm during this time, Cuba remained largely unaffected by "Zoom fatigue." Cuban students relied on televised classes and occasional phone conversations with their teachers if they were fortunate. University students received assignments through WhatsApp, often with little additional support, and were expected to work independently. In such circumstances, el paquete continued to offer Cubans a connection to the broader world.

Despite (or perhaps because of) the economic uncertainties and with ample idle time, people yearned for entertainment, leading to an increased demand for el paquete and placing a greater burden on paqueteros. Yasmani, thirty-five, who delivered the paquete in Ciego de Ávila, vividly recalled the surging demand he encountered:

> During the height of the pandemic, it was forbidden to be outside after one o'clock in the afternoon. Many people were confined to their homes with their families, unable to go out due to either being suspected cases or testing positive. They were in dire need of cartoons for their little children, soap operas for their parents, anything that could keep everyone at home calm and entertained.

During that period, Yasmani experienced a significant increase in his customer base. While he did lose a few customers who could no longer afford the paquete or wanted less frequent deliveries, he received numerous calls from new clients who wanted to subscribe to his service. Additionally, he took on the subscribers of other paqueteros who had stopped working due to fears of contracting the virus. Recognizing that many of his customers were elderly and at higher risk, Yasmani implemented "contactless delivery" measures. He would wrap the hard drives in plastic, disinfect them with a chlorine solution, and then leave them on his customers' windowsills. As the pandemic continued, Yasmani and other paqueteros in my research became essential frontline workers, ensuring their clients' access to global media culture and providing a much-needed source of entertainment during the lockdowns.

El Paquete's Impact on Citizen-State Relations

As a vernacular infrastructure that compensates for the limitations of state-run projects of infrastructural provision, el paquete semanal has fundamentally transformed the relations between the citizenry and the state. It has created an alternative media ecosystem that provides access to content excluded from state media, promotes a consumer logic that challenges the values of the paternalistic socialist state, and establishes a parallel economic space that grants greater independence to the emerging private sector. Throughout this chapter, I have emphasized the economic opportunities that el paquete has opened up for many Cubans working as paqueteros, running copy stores, or utilizing it as a platform to promote their small businesses, music, art, or journalism. Moreover, it serves as an independent platform for groups like evangelical churches, LGBTIQ+ activists, or video bloggers to disseminate their content and messages across the island, bypassing state media channels that may not provide them with representation.

However, matrices and copy store owners are cautious about being seen as competitors to the official state media monopoly. They ensure that the content they distribute is not critical of the government and actively practice self-censorship by excluding pornographic or politically sensitive material that could attract state scrutiny. In some cases, they even include state-produced content, such as the *Gaceta Oficial*, which provides information about legal changes. El paquete, therefore, cannot fully substitute for an independent public sphere, and the freedoms it offers are primarily economic rather than political.

The state is well aware of the significant economic and cultural impact of this distribution network, not least through numerous government-funded research projects conducted by social scientists at institutions like the Instituto Cubano de Investigación Cultural Juan Marinello or the Facultad de Communicacíon at the University of Havana. The government deliberately keeps el paquete in an extralegal limbo, neither restricting its sale nor establishing a legal framework for it. However, there have been instances where the government intervened in the distribution of el paquete for political reasons. In July 2018, when an episode of *El Señor de los Cielos*, a series produced by the American Spanish-language terrestrial television network Telemundo, portrayed Raúl Castro and General Ramiro Valdés as the main drug traffickers in the Caribbean, State Security contacted the two major casas matrices to halt the distribution of the highly popular series, and both complied. Odisea publicly announced in their Paquete edition that they would no longer broadcast the "disrespectful" show by issuing a statement in the form of a PDF file. On the other hand, Omega edited out the problematic scenes and released a version without the offensive content.

Cuban president Miguel Díaz-Canel has made references to el paquete in several speeches and interviews, acknowledging that he does not oppose it in principle but criticizing the values it promotes and the cultural behavior it encourages. However, the government, recognizing that el paquete fulfills certain significant needs of the Cuban population and serves as a source of distraction and entertainment, tolerates its existence. This relatively lenient stance by the state is a direct result of the economic and social reforms initiated by Raúl Castro since 2010. These reforms involved layoffs of state workers, reductions in social spending, and a notable expansion of the private sector, in which el paquete plays a vital role as a distribution and communication platform. In addition to introducing market mechanisms, these reforms have redefined the roles of state institutions, economic actors, and the general population, resulting, as highlighted by Hoffmann (2016), in a de-ideologization of Cuban politics. While socialist principles are still upheld and promoted, they are no longer strictly enforced.

This de-ideologization has led the state to accept the loss of its control over the cultural consumption of the population. State media have adjusted their strategies accordingly, attempting to compete with el paquete by shedding their previous resentment and endeavoring to emulate its successful formula. It is possible that Cuban authorities have heeded the research conducted by institutions such as the ICIC Juan Marinello and the University of Havana, which have consistently demonstrated that the shift in cultural consumption from official media to independent alternatives is primarily driven by the younger generation. This younger demographic strongly rejects the ideologically charged content found in state programming, as highlighted by studies conducted by Yecenia Brito Chávez (2014), José Raúl Concepción Llanes (2015), and Isabel Echemendía Pérez (2015).

Despite state television remaining the primary source of information for the majority of the Cuban population, complaints about the limited choice and quality of content are widespread. Recognizing this dissatisfaction, the state has initiated the development of official alternatives to el paquete that attempt to replicate its offerings to some extent. One such alternative is Mochila (back-pack), a curated collection of data distributed by the government since 2017. Mochila can be copied for a small fee at government-operated cybercafés and youth computing clubs across the country.

Mochila typically includes a significant amount of official government information and propaganda, akin to what is found in the national mass media. It encompasses educational software, exercises for national university entrance exams, as well as PDF files featuring poems by the national hero José Martí or reflections by Fidel Castro. One folder within Mochila, titled "De mi terruño" (From my homeland), features productions from regional television and municipal telecenters throughout Cuba. However, this more official and educational content is interspersed with materials that closely resemble those found in the private paquete. For instance, the folders "A jugar" (Let's play), "Aplicaciones" (Applications), and "Utilisimo" (Useful) provide a comparable range of video games, software, and mobile applications, along with a selection of YouTube tutorials on makeup, cooking, and fitness that closely align with what can be found in el paquete. Likewise, the telenovelas and Korean dramas accessible in Mochila are virtually indistinguishable from the content offered by private paquete distributors.[4]

Throughout its existence, the state-run distribution platform Mochila has failed to achieve the cultural influence enjoyed by el paquete. Few Cubans I encountered during my research consumed Mochila regularly, and according to their accounts, this had less to do with its contents and more to do with the less organized distribution system. While el paquete could be conveniently delivered to one's home, Mochila required individuals to visit a Joven Club de Computación or Sala de Navegación to make copies on often occupied, outdated machines.

On the other hand, the state has had more success with the creation of Picta, a streaming platform developed by the University of Information Sciences (UCI) that bears resemblance to Netflix. Picta allows viewers to stream and download not only Cuban content but also international series, live soccer matches, and Hollywood movies. The significant advantage of Picta is that it operates as a Cuban website and only consumes customers' national data plans, which are much more affordable. Additionally, national data packages are often included as bonuses with top-ups from abroad, which many Cubans receive from their family members. For the average cost of a paquete, one can enjoy approximately twenty hours of surfing on Picta. Lazaro, a paquetero from Centro Habana, expressed the sentiment that Picta is providing real competition: "Cubans always look for the most economical solutions, and by raising the price of el paquete, we have lost customers."

Picta is frequently promoted on the online government news outlet *Cubadebate*, particularly through its Canal USB section dedicated to culture and technology, which aims to attract a younger audience that may be less interested in traditional state newspapers such as Granma or Juventud Rebelde. Canal USB often features film reviews, including works like the 2021 Oscar winner *Nomadland*, accompanied by a link to watch the movie on Picta. While the government's strategy with Mochila may have been to entice people to access its educational materials by including international entertainment content as a Trojan horse, Picta primarily focuses on global entertainment, presenting a more serious alternative to el paquete. However, this comes at the expense of almost complete assimilation to it.

Conclusion

With the introduction of Mochila and Picta, as well as the increasing inclusion of international entertainment programming on official television stations, the Cuban state seems to be adapting to the evolving media landscape by emulating its independent competitors. By tolerating the rise of private initiatives like el paquete, the state has relinquished much of its control over media distribution on the island. Rather than regulating these private initiatives, the paternalistic socialist state is creating official alternatives, ultimately succumbing to the capitalist attention economy. The tremendous success of el paquete has accustomed Cubans to the constant availability of global media productions. While the government can still ensure that politically oppositional content is not disseminated through this alternative infrastructure, it seems to have given up on trying to control the social values it conveys. In a way, the state itself is now participating in the materialistic cultural behavior that President Díaz-Canel criticized el paquete for, as it feels compelled to utilize capitalist entertainment formats to maintain its relevance as an information provider.

Therefore, el paquete represents a crucial arena where the relationship between citizens and the state is currently being redefined. It has brought about greater economic opportunities and freedoms for the Cuban people, as it provides an infrastructure for accessing a wide range of global media, expressing individuality through consumption, distributing personal media content, promoting small businesses, and earning income through its distribution. All of these developments have marked the initial stages of the gradual erosion of the state's media monopoly, which holds immense social and political implications that will be explored in subsequent chapters.

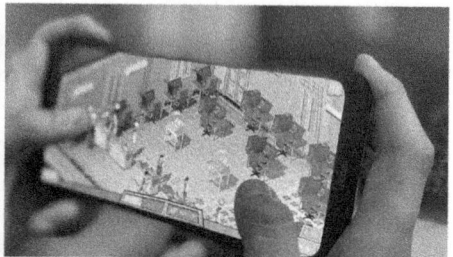

PakeTown (2021)

Documentary mobile video game for Android/iOS
BY STEFFEN KÖHN, NESTOR SIRÉ, AND CONWIRO

www.da.gd/paquete

Project lead: Rafael Martínez
Technical lead: Javier Domínguez
3D lead artist: Orestes Suarez
Unity developer: Aniel Ramos
2D artist: Lázaro Piñol
Back-end lead: Alejandro González
Community manager: Harbin Betancourt

PakeTown is an ethnographic mobile game that depicts the history of alternative Cuban offline media distribution. It focuses on the period from the 1970s, when paperback novels became the first entertainment materials distributed outside of government control after the Revolution, up until the present day with the hard-drive-based sneakernet el paquete semanal. In the game, players assume the role of entrepreneurs in the informal Cuban media sector who operate a media rental business from their homes. The game consists of

FIGURE 2.2. Screenshots from *PakeTown* and images from the promotional campaign designed by X-Dot, one of the first PR agencies to specialize in offline ads within el paquete.

five levels, each representing a different decade. Players start by renting out romance and western novels to earn money, which they can later invest in VHS recorders, cassettes, and illegal satellite antennas that allow them to grab U.S. and Latin American TV programming. As the game progresses, players must adapt their businesses to changing media formats such as DVDs and eventually hard drives, which became more prevalent with the increasing availability of computers on the island during the 2000s.

Aesthetically and conceptually, *PakeTown* draws inspiration from the genre of business simulation or tycoon games. Nestor and I came up with the idea for the game when we noticed the high popularity and consumption of business simulation games like *Pizza Syndicate*, *FarmVille*, and *Resort Tycoon* within the paquete. We found it intriguing how these games thrived despite (or maybe because of) the many restrictions on the Cuban private sector. In order to create an accurate portrayal of the history of informal Cuban media circulation networks, we meticulously documented the changing media contents, techno-logical devices, and organizational strategies through biographical interviews with owners of informal video banks and copy stores as well as paqueteros.

PakeTown was developed by one of the island's first independent game studios, ConWiro. At the time we approached them, they had already com-pleted four video games, including one commissioned work to promote a Cuban dating app. ConWiro's team consisted of several programmers who had learned their craft as SNET members, giving them a natural connection to the game's content and topic. Unlike our previous projects, which we primarily premiered at American or European festivals or institutions due to equipment limitations in Cuba, we aimed to create a work specifically for the Cuban mar-ket that would cater to the preferences of local audiences. Therefore, we made *PakeTown* available for free through the paquete, the very medium it portrays. After lengthy negotiations with state institutions, ConWiro also succeeded in having the game included in Apklis, the state-run national app store that replaces the restricted Google Play and Apple App Store. Apklis features local Cuban applications as well as pirated essential international ones. International audiences can access *PakeTown* on both Apple's and Google's app stores in Spanish, English, and German (as the production of the game was funded by the German embassy in Cuba and the Goethe Institute).

To promote the game in Cuba, we organized public presentations at Ha-vana Espacios Creativos and a Joven Club de Computación in Camagüey. Obtaining permission to present the game turned out to be more challeng-ing than we initially anticipated. For example, staff members at Havana's Palacio Central de Computacíon, the headquarters of the government-run youth computing clubs, expressed concerns about the political implications of showcasing a game centered around an extralegal phenomenon. Outside the capital, these concerns were less pronounced, highlighting the varying degrees of discretion decision makers have within the hierarchical state structures.

Collaborating with ConWiro over a span of more than two years, with the goal of creating a successful product for local audiences, provided invaluable insights into how a private company in the underdeveloped Cuban technological sector navigated the complex technical and political limitations during a time of profound social change in Cuba. Our interventionist research approach allowed us to directly observe how Rafael, who is in charge of business operations, and Javier and his team, who are responsible for the technical side, addressed both internal and external challenges. Internally, they faced restricted Internet access, bureaucratic barriers, limited access to venture funds, and a weak legal framework for private businesses. Externally, they encountered additional limitations imposed by the U.S. embargo, such as being excluded from international payment traffic and lacking access to essential software and programming tools. For example, they could access the cross-platform game engine Unity, in which *PakeTown* was programmed, only via VPN and had to rely on a contact in the United States to upload their games to the global mobile app stores. Throughout our research, we closely followed ConWiro's journey as they worked toward legalizing their business. The state had stopped issuing licenses for private software developers, but they found a way to register their company as an independent studio of audiovisual production with the national Instituto Cubano del Arte e Industria Cinematográficos (ICAIC; Cuban Institute of Cinematographic Art and Industry). Once officially recognized, they were able to create links with state institutions to launch their products and earn revenue. Notably, they achieved a significant milestone by becoming the first Cuban private company authorized to charge micropayments via SMS within apps on the official national app store. While maintaining *PakeTown* as a free-to-play game, Nestor and I made the decision to permit ConWiro to introduce micropayments within the game. These micropayments provided players with the option to progress at an accelerated pace, while also granting the company the opportunity to further enhance and expand the game in line with their business expertise.

The linked data package includes the game in both Android and iOS formats, PR materials used to promote the game (designed by X-Dot, one of the first PR agencies specializing in offline ads within el paquete), installation views from its exhibition at the Moesgaard Museum in Aarhus (for which we re-created a Cuban data copy shop), and photos from the public presentations of the game in Cuba.

Cuban gamers at a DOTA 2 LAN party-02.jpg

A biquad WiFi antenna made from a USB WiFi.jpg

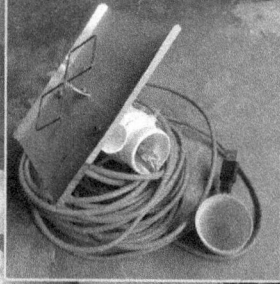
PC of a typical SNET gamer .jpg

Anti-theft and weather protection cages for the network's WiFi antennas -02.jpg

SNET P4...

Server of CG Net in Camaguey .jpg

CHAPTER 3
Networks

"And this router here connects to Republic of Gamers, the main game hosting pillar," Erick, a warm and open-hearted man in his mid-forties, gestured toward the east, casting his gaze across Havana Bay to a point visible only to him. Our vantage point was the rooftop of a large socialist building, situated just two blocks away from Parque Trillo in the heart of Centro Havana. Adjacent to us stood a tall mast adorned with multiple Wi-Fi antennas, its presence towering over the surrounding houses. This prominent structure served as the conduit, extending connectivity to all parts of the city, forming the central backbone of SNET—the locally cultivated computer network that seamlessly traversed the entirety of Havana. A self-professed tech enthusiast and skilled jewelry maker with hands that carried the subtle marks of years spent refining his craft, Erick took pride in his role as the administrator of the SNET backbone—a crucial component of the network infrastructure that facilitated a significant portion of its traffic. He eagerly demonstrated the security measures he had implemented with care. A watchful security camera monitored the top floor of the stairwell, while a protective ring adorned with narrow blades acted as a deterrent, dissuading potential trespassers from attempting to scale the tower. Descending to the lower floors, we found ourselves in Erick's apartment, where his son, Erick Junior, guided us to his room. A cluttered desk housed a computer and an assortment of electronics and equipment, including USB hubs, headphones, and cables. As an esteemed Minecraft administrator within the SNET community, Erick Junior enthusiastically shared a sneak peek into the latest mods and cracked games that

FIGURE 3.1. Havana's citywide gamer network SNET grew out of neighborhood LAN parties and relies on DIY equipment. Its physical infrastructure consists of miles of Ethernet cables running through streets and across balconies, along with thousands of Wi-Fi antennas mounted on rooftops, network switches, and servers.

would soon be accessible through the network, exuding an aura of secrecy and excitement.

Grassroots computer networks like SNET have emerged as another bottom-up response to the limitations of Internet access in Cuba, capturing the dedication and ingenuity of passionate technology geeks across the country. These networks provide an avenue for users to engage in various activities such as multiplayer video gaming, chat, messaging, forum discussions, file sharing, and even website hosting. Among these networks, Havana's SNET stands as the largest and most prominent. It has grown organically from the interconnection of numerous neighborhood-based LANs and is believed to be the world's largest community network that operates in complete isolation from the global Internet. SNET has successfully brought together tens of thousands of users, fostering a vibrant and interconnected digital community within the city (Dye et al. 2019; Pujol et al. 2017). Its material base consists of miles of Ethernet cables running across streets or balconies, Wi-Fi antennas mounted on poles on rooftops, and servers and network switches operated by an army of volunteer node administrators. It relies on a network of thousands of participants who collaboratively create, operate, and maintain its hardware and software infrastructure. This underlying social structure is crucial to the network's evolution and persistence, as its members collectively provide the funding for its technological basis, connections between individual nodes are mediated by personal relationships, and admins and users must work together to ensure its survival in a complex social and political context.

The story of SNET is instructive as it illuminates both the possibilities and the limitations of such private initiatives that create new forms of autonomous self-organization in an authoritarian environment. Compared to el paquete semanal, which primarily provides Cubans with access to entertainment and is therefore tolerated by the state, the much smaller phenomenon of grassroots computer networks has provoked a more ambivalent response from the government. Therefore, SNET makers were forced to constantly adapt to shifting technical, political, and social frameworks.

While SNET effectively became illegal in August 2019, when the Cuban government introduced new regulation on the use of Wi-Fi, some of its admins successfully negotiated its integration into Tinored, the official nationwide intranet that connects the state's network of Joven Clubs de Computación, which secured the survival of many of its offerings but resulted in the loss of its autonomy. As I show, the protests and negotiations for SNET's survival in many ways anticipated later conflicts that had major political implications.

SNET further provides a fascinating case to better understand the dynamics of people-embedded infrastructures in politically restrictive contexts. Researchers oftentimes have celebrated the creative, flexible ways in which such informal networks step in when official infrastructures fail to deliver (Larkin 2004; Schwenkel 2015; Storey 2021; Victor 2019). In contrast, I want to highlight the "dark sides" of such human infrastructures, the immense stresses that often rest on their underlying social networks, and the conflicts or breakdowns that inevitably arise when the people involved in them have conflicting ideas, are involved in power struggles, or react differently to external pressures. By calling attention to their inherent fragility and the consequences of their breakdown, I demonstrate how attempts to repair or renegotiate the interpersonal relationships that support these human infrastructures fundamentally shape the ways in which they are experienced and how this impacts their generative potentials.

Moments of material breakdown have been identified as central to the basic operation of infrastructures in many seminal writings on the concept, such as in Susan Leigh Star and Karen Ruhleder's (1996) definition of infrastructures as invisible until breakdown, Brian Larkin's (2008) insistence on technological collapse and repair as constitutive elements that form both people's relationship with infrastructures and the imaginaries these produce in the Global South, and Steven J. Jackson's (2014, 221) call for "broken world thinking" that accounts for the fragility of the technical worlds around us. For Cymene Howe and colleagues (2016), ruin is a central paradox of infrastructure, as it is doomed to degenerate despite its generative qualities, deteriorating in spite of its inherent promise of progress.

I expand these materialist perspectives on infrastructural fragility by analyzing similar ambiguities in people-embedded infrastructures such as SNET, where practices of collaborative creation and maintenance create new forms of sociability, belonging, and solidarity among previously unconnected people but can also provoke internal conflict, require complex negotiations, or even lead to the breakdown of the social ties on which they depend. SNET is a particularly revealing case in point because the collapse of relationships between high-level admins over their competing ideas of what SNET should be—a horizontal community network or a more hierarchical structure with commercial elements—ultimately led to its physical separation.[1] This eventually made it easier for the state to end the network's anonymity and independence, as users and administrators could not agree on organizing a unified response to the new regulations.

SNET's Origins

IT-savvy Cubans began tinkering with the technological components necessary to form computer networks when these slowly began to arrive in the
country in the early 2000s. As Cuban law (with its 1992 Decree 171) prohibited the private use of the radio spectrum without authorization from the
state (initially to ban private ham radio operators), wireless technology was
available only on the black market. Around the same time, the proliferation
of desktop computers on the island made private LAN parties popular, where
young gamers would carry their computers to a friend's house and join them
via Ethernet cables to play multiplayer video games like StarCraft or Defense
of the Ancients (DotA). To avoid the tiresome moving of gear, some of these
gamers eventually began to string cables through windows and across balconies or even streets to permanently connect to friends in their vicinity. These
basic computer networks then grew with the increased availability of wireless
access bridges (locally called "nanos" after a popular model, Ubiquiti's Nano-
Station) so that soon whole neighborhoods became interlinked. Ernesto, aged
thirty-two, a former admin and early SNET member who currently works as
a software developer in the state sector, explained how this technology generated new connections between hitherto unconnected people:

> Someone from the neighborhood got a switch, we got network cable, so
> we managed to start assembling a network infrastructure. At first it was
> only in my building, in my neighborhood, but soon after we had started,
> someone said, "Look, we have something similar a few blocks away. Let's
> join the two networks." We could not do it wired because it was more than
> a hundred meters and a lot of signal was lost, so we decided, "Well, let's
> do it wireless," and so it started to grow in such an organic way that I don't
> think anybody saw where it was going.

The practice became so popular that by around 2009 almost every *municipio* of Havana had its proper gaming network, and soon the idea took shape
to connect these individual nodes to form a much larger network. Around
2011, groups of nodes began meeting regularly to explore how the growing
network could be cooperatively sustained, with new forms of collective action
and new political subjectivities emerging along the way. In a tightly controlled
society based on centralized economic planning, the members of these nodes
established new forms of community organization and decision-making. A
pyramidal structure was created in which smaller local area networks (linked

by Ethernet cables and network switches) as subnodes came to be connected to larger nodes via wireless bridges. These nodes were managed by local administrators who also provided technical support for the hundreds of users connected through them. Individual nodes, in turn, became linked to one of nine regional pillars (grouped by geographical proximity and run by general administrators) via long-range directional Wi-Fi devices. These pillars were joined by fixed wireless links, with each pillar peering with at least two other pillars and the central backbone. With the growth of the network, more and more people got together to communally develop services that quickly surpassed the net's early focus on gaming, such as moderated discussion forums (first around video games, then about all kinds of different topics), file-sharing platforms, blogs, mirrors of websites from the World Wide Web, and even homemade social network sites. The most important of these services were hosted at the pillar level (Republic of Gamers [RoG], for example, is the pillar that hosts the servers of the most popular games), while the individual nodes provided FTP servers for the sharing of local content. Such community networks developed in many Cuban cities. Ernesto describes,

> Holguín has its own variant; Camagüey has its own variant. Almost all the provincial capitals have their own SNET variants, all with their very unique subculture. For example, here in Havana we had a group that produced electronic music, we had our forum, we had our nonsense, we had a channel to broadcast live audio on TeamSpeak. And then that same community existed in Camagüey. But the goal in Camagüey I remember was more commercial, more trying to get gigs in clubs while the community in Havana was more like, "Listen to this noise I made with my computer and tell me if you like it." . . . These local networks often exist because of content that is of interest to a particular community.

SNET thus evolved in the form of ever-widening circles whose diameter increased as their makers' access to technology improved, allowing for greater connectivity. Likewise, its governance structure evolved in parallel to its technical evolution, with each regional node or pillar deciding on roles and responsibilities among themselves. At the top of these hierarchies were the central administrators (on the pillar level), followed by admins of local nodes. These were usually people who had the most profound technical knowledge and/or had invested the most in network technology and servers and, therefore, had gained the respect of the community (Rodríguez Fernández 2019). They were usually supported by a technical team that operated on a voluntary basis, mainly for the social status this gave them among members, firsthand access to information, or some influence on the structural or normative modifications of the net.

Aware that the revolutionary Cuban state closely monitors all attempts at autonomous social organization, SNET makers collectively agreed on a set of common rules that every user of the network had to follow. To avoid government repression, they proactively implemented strict policies banning all content that authorities might deem provocative, such as politics or pornography. Users happily accepted such self-censorship because the vast majority of them were not interested in SNET as a potential platform for free expression or political debate but rather approached it in more practical terms as an infrastructure that allowed them networked gaming and file sharing and that emulated some aspects of the Internet, such as social media, services otherwise not available to them. When the state finally introduced Wi-Fi in public parks and the Nauta Hogar home service from 2015, admins made sure that no node bridged to the Internet, which would have been illegal, as SNET would then have competed with the state telecommunications monopolist ETECSA. Social misbehaviors such as trolling and online vandalism were, of course, also sanctioned. Wrongdoings were classified from light to very severe, and the corresponding disciplinary measures ranged from brief disconnection to permanent expulsion. For the practical reason of not overloading the network, the admins also set fixed times for playing games and copying data (daytime was play time and file transfer was allowed only from three o'clock in the morning to noon).

From a legal perspective, the network for a long time existed in limbo. Even though it was made up of Wi-Fi equipment that (before 2019) was prohibited from being brought into the country, the SNET community otherwise took pains to avoid any confrontation with the government. While private wireless networks technically also were banned, admins felt that much of the legislation (stemming from the 1990s) was obsolete, as it was not yet imaginable back then that a customary mobile phone would one day be able to serve as a wireless access point. SNET users therefore saw it as a positive sign of official recognition when in 2016 the government news website *Cubadebate* published a flattering portrait of the community, complete with a schematic map of the network, fully disclosed names of admins, and even a group photo. A year later, a similarly positive article even followed in the national newspaper *Granma*, the official organ of the Communist Party of Cuba. Such signs of cautious tolerance led to a widespread feeling among users that they finally had received some sort of governmental acceptance.

SNET's Member Base

SNET members were mostly male youth between the ages of fifteen and about thirty. A typical SNET user's career peaked before university, when students still have lots of free time to invest in gaming and the kind of voluntary work that the maintenance of the network required. Many older users and admins had fond memories of this time in their lives and described that such intense dedication to the network typically led to two outcomes: people either managed to significantly reduce the time they invested in SNET when they reached university (yet often built on what they had learned by studying informatics, computer science, or a similar career) or became so hooked that SNET became a central part of their lives. SNET admins tended to be a bit older and had risen in the network hierarchy either because of their technical skills or because they owned the equipment through which many users were connected. In sociological terms, SNET users and admins were mostly middle to upper-middle class, meaning they (or their parents) had remittance-sending families abroad, a well-paying job in the private sector, or a secure state job that came with "freebies" such as access to computer equipment or the Internet. Although I met admins of important nodes who lived in run-down housing and worked on very old machines, many others had the means to regularly acquire new gear through the black market or relatives in the United States. Further, SNET's member base was notably "whiter" than the general population, as phenotypically white Cubans are much more likely to have access to the foreign currency or remittances necessary to buy computers and tech equipment.

While there were women who actively contributed to SNET as forum moderators, radio show hosts, or online support personnel, they were markedly absent from all administrative positions that held some influence or power. As Greta, a thirty-year-old journalism graduate and member of one of the World of Warcraft developer crews, remarked,

> When people chat with me on TeamSpeak, they always think that I'm a man. They call me uncle, brother, dude, are you there, I need help. And at some point, I no longer said, "Hey, I'm a girl or I'm a woman." I just answered their requests because I had no choice, because it's too much. People are not adapted to women in administrative positions, and they are not adapted to women who develop. It's quite shocking.

Such gendered dynamics of power are prevalent in many global tech and gaming cultures that are marked by the exclusion of women and sometimes even outright misogyny (Vickery and Everbach 2018), and SNET is no exception. Moreover, in Cuba, the pervasive ideology of self-making, which will be discussed in the next section, is strongly associated with male spheres of work or economic activity (Härkönen 2016, 14), which further reinforced the network's lack of diversity.

Infrastructuring as a Cultural Ideal

Due to its organic evolution, SNET never had a solid structure and continuously faced many problems: a material basis insufficient for its expeditious growth, occasional crackdowns by local authorities (such as neighborhood "Committees for the Defense of the Revolution" checking for antennas on rooftops in order to keep Cubans from receiving the U.S. propaganda broadcaster Radio Televisión Martí), and disagreements and rivalries among administrators. A key reason for its resilience was its decidedly distributed infrastructure, which guaranteed that the network continued to work even when parts of it became unavailable. As a mesh network, SNET's individual nodes connected directly, dynamically, and nonhierarchically to as many other nodes as possible. The lack of dependency on one node allowed for every node to participate in the relay of information. This distributed infrastructure was deliberately designed so that it could easily be modified to adapt to changing conditions. Modification was a central task for SNET makers, as the video games that were played over the network had to be "modded" to function in Cuba without a software license and without connecting to the Internet, and its hardware and software base constantly had to be refashioned. The concept of "modding" has its origins in gaming cultures and describes alterations by fan programmers that change one or more aspects of a video game (Postigo 2007). It is a communal practice that entails an ethos of collaboration my research participants frequently evoked. I use the rubric of modding as a prism to understand how SNET members not only manipulated their network's software and physical infrastructure but also made and remade social connections and enacted agency in political processes.

As a practice that represents an almost infinite combinability of ideas, materials, and applications and demonstrates makers' ingenious aptitude for innovative responses (Jungnickel 2014, 93), modding is an indispensable part of the Cuban everyday practice of resolver, the pervasive cultural ideology that denotes the inventive overcoming of great obstacles with minimal resources by repurposing what is available (Dye 2019).[2] Resolver became a necessity when the Soviet Union collapsed in 1991 and Cuba entered its Special Period, which saw unprecedented shortages of everyday items. The practice continues to be essential in the island's still-prevailing economy of scarcity. It is part and parcel of *la lucha*, the daily struggle for survival in an ongoing economic crisis. Kathy Powell (2008, 181) asserts that resolver is a collective effort because its ethos depends on community solidarity, relationships of trust, cooperation, and social obligations of reciprocity that sustain networks of relationships among family, neighborhood, and community members and that become particularly critical resources in times of need. She shows that part of the dialectic of resolver is that this communal solidarity is achieved at the cost of immense strain on the very social relations that constitute that solidarity, thus constantly threatening the strong bonds on which it is based. This burden, which also rested on the social relationships that formed the "back end" of SNET's material infrastructure, became clearly visible during my research. Admins often complained about how they were dependent on people with whom they did not share a common vision, whose way of doing things they disliked, or with whom they had conflicts over money, but with whom they still had to collaborate because of SNET's limited resources.

In theoretical terms, resolver can be understood as a form of "infrastructuring" (Star and Bowker 2002), as part of the infrastructural and articulation work (Star 1999) needed to keep networks like SNET alive. As Star describes, such work is being learned as part of the membership in a community of practice and constitutes a set of cultural competences. Challenging Star's argument about the invisibility of infrastructure and the work that keeps it running (which she argues becomes apparent only when it collapses), Larkin (2013) insists that infrastructures are often very present in people's lives, mobilizing affects and producing strong feelings of longing and pride, desire and frustration. Such conflicting emotions indeed were present among my research participants. Whenever SNET makers spoke about their achievements in collectively reinventing, modding, and maintaining the network, they expressed both pride in their ingenuity and frustration about the large investments necessary in terms of work, time, and reciprocity. Instead of constantly being dependent on others, they often stated that they would prefer the supposedly smoothly running consumer Internet access available in other countries. This serves as an important reminder that celebrating resolver and other sophisticated local forms of modding, resourcefulness, and invention also entails the risk of fetishizing practices created out of pure necessity by those excluded from consumer capitalism (Dye et al. 2019).[3] In what follows,

I discuss how my research participants used various practices of modding to deal with the fragility of the material and social infrastructures on which SNET was based.

Modding the Software Infrastructure

In its heyday, SNET hosted hundreds of websites, among them discussion forums on every imaginable topic, Craigslist-style classified advertisements platforms, several Facebook alternatives (such as Sígueme, or "Follow me"), mirrors of Internet websites (such as Wikipedia), and even various search engines to locate content. Most of these services relied on modified software that had been appropriated for use within SNET. The discussion boards, for example, were based on phpBB, a free, open-source flat-forum bulletin board software. Video games were provided via pirated versions of digital distribution platforms, such as Valve's Steam and Blizzard's BattleNet, that had been heavily modded so that all games ran free of charge and never went online.[4] They were also regularly patched manually to ensure that they always ran on the latest versions. As Alexander Knorr (2009) describes in his ethnography of a global community of game modders, modding is a highly transnational collective practice that hinges on an ethos of participatory culture. This is of course also true for Cuba, where most active game modders were members of the RoG pillar. Despite the limited Internet access on the island, RoG's developer team had managed to sustain relations with the global modding scene. RoG's central administrator, DaVinci, through personal contacts even was able to convince Blizzard, the company behind World of Warcraft, to give the software to the Cuban community, allowing them to adapt the game to the specific needs of their network.

The central interface that users utilized to navigate SNET also was a repurposed piece of software. TeamSpeak (TS), a voice-conferencing software that allows users to communicate with each other via voice and text over the Internet or a LAN, had developed into the preferred platform for the users of Cuba's community networks to connect to other users but also to forge their digital identities. While TS is firmly rooted in gaming culture (being designed for gamers who can use the software to communicate with other players in the same team in a multiplayer video game), it was employed in SNET as a central organizational tool for its many features, customizability, and low system requirements. For most users, TS was the first thing they opened when they entered SNET. It was organized into a slew of channels and subchannels. Each pillar had its own TS server that functioned as the main communication platform within SNET, but with the growth of the network, pillars allowed nodes and subnodes to also host their own TS servers. The entry channel of each TS is the lobby in which all users are listed with their nicknames. Admins used the lobby as a directory that provided links to the various websites and services of a node or pillar.[5] They also employed it to transmit messages by moving all connected users to this channel when they needed to make a public announcement. Users could use TS to contact all the other network members via voice chat, send private or public messages, and send and receive data files. They could also create their own temporary channels on various topics with the approval of their admin. Admins and users made apt use of the software's high customizability and appropriated some of its features in order to make it work more like a social media network or a content-streaming platform. They, for example, delighted in using custom skins and themes and created channels for the broadcast of live DJ sets or self-made radio shows. As the TS lobby served as a central user database for each node or pillar, users invested a lot of care in their profiles (which were their prime digital representations within the network) by uploading an avatar image and choosing a nickname that was often designed in the aesthetics of ASCII art (for example, #!MαĴaÐɛřΘ, ⚡★⚡FŁΔ$Ħ⚡★⚡, or N3PH3L1M). Each profile provided additional information about the user through icons that the admin and technical team of each TS individually created. These icons appeared behind the user's nickname and let them self-identify, for example, as a fan of a particular game or football team, but also provided information about their standing in the network's hierarchy (e.g., as being part of a technical or developer team or a forum moderator). Admins could also attach icons to user nicknames to publicly flag them if they broke any rules.

Richard Rogers (2019) has periodized the history of the global Internet as a succession of three distinct logics of navigating it: first, the open web that was traversed by surfing and organized by web directories (such as DMOZ or Yahoo); second, the amateur Web 2.0 Internet accessed by search engines (like AltaVista or Google); and third, our current centralized, capitalist net of closed platforms and its logics of the feed in which we passively scroll through content that has been tailored to our algorithmically determined interests. Instead of the online directory, the search engine, or the feed, SNET and other community networks around the country made TS their central interface and were therefore organized by a logic of the social that prioritized communication and interaction between all users of the network. As we will now see, such a system, which promotes direct contact between all its members, was a prerequisite for maintaining the network infrastructure.

Modding the Physical Infrastructure

The physical limitations of Wi-Fi, the tropical climate, and a complex political context (with the state banning the import of wireless antennas and the U.S. embargo raising the cost of networking and computing equipment) all put a lot of strain on the physical infrastructure of Cuba's community networks. These factors contributed to the networks' inherent fragility, which required constant acts of resolver. Wi-Fi is constrained by the use of the 2.4 GHz and 5 GHz spectrums, waves that by nature don't travel very far and are easily absorbed by trees, weather, or the built environment. Wi-Fi antennas therefore need line-of-sight links, which means that they often have to be mounted on poles on top of buildings to ensure that there is no interference. They also must be protected from the extreme tropical climate because technological devices placed outside can easily suffer from sunburn or become fried by a lightning storm during the rainy season (Dye et al. 2019). SNET creators therefore had to mod found objects such as plastic containers to create housings for gear, find suitable poles, and construct elaborate antitheft structures (such as spikes or even electrified covers) to secure their costly equipment.

Alvaro, in his early fifties and a law graduate who now works as a private taxi driver, used to administer a node in Alamar. He remembered that at the beginning some nodes, due to the impossibility of finding manufactured antennas, even relied on self-made directional Wi-Fi antennas that were built from cheap wireless network interface controllers, some copper wire, and scrap metal. Even though they were fully functional, users replaced them whenever a professional antenna became available on the black market as the DIY models (locally called *biquads*) just weren't fast enough and created latency during gaming.

The network was therefore shaped by what members could contribute in terms of materials, resources, and skills. It was designed so that it remained functional and rerouted traffic if particular nodes or connections failed because of disruptive nonhuman (such as rainstorms) or human elements (such as thieves). Camilo, a medical student from Centro Habana, described how in his subnode, users always temporarily disconnected their equipment and brought Ethernet cables (which are not meant for outdoor use) inside during a storm because once a lightning bolt had killed their local nano and it took them months to be able to replace it. Hence, users had learned to interact with such agents and forces outside of their control to make their network more resilient. TS, the communication platform that connected all participants in the network, ensured that users could immediately report all problems with the physical infrastructure so admins (depending on the importance of the connection point that broke down) could determine how to react. However, as I will now show, continuous breakdown affected not only SNET's physical infrastructure but also the social relationships that made up the network that carried the risk of collapse.

Modding Social Relations

SNET's human infrastructure relied on enduring social ties between its members. Because almost every aspect of network configuration required direct contact between users and administrators, communal construction and maintenance necessitated a lot of voluntary work, and members often organized meetings and events, SNET generated a tight-knit community with a strong sense of belonging. Yet SNET users depended on people not only inside but also outside the network to make connection. Getting permission to mount an antenna on a spot with good reception or running cable over a balcony frequently required the consent or goodwill of neighbors, parents, or partners. When equipment needed to be repaired or replaced, this often led to lengthy renegotiations as a neighbor's property needed to be accessed again or household savings had to be invested in new cables or antennas. The supporting relationships necessary to maintain this local network oftentimes transgressed national borders, as friends or relatives outside of Cuba frequently provided access to Wi-Fi equipment, the resources to acquire it, or vital information on how to set up or mod hardware or software.

Echoing Larkin's (2013, 333) observation that collectively constructing infrastructure is often experienced as deeply emotional and fulfilling by its makers, research participants univocally described how SNET had enriched their lives, how online acquaintance often seamlessly translated to friendship, or how the continuous exchange in forums or during meetings had taught them invaluable technical skills and created a strong community spirit. Admins, who all stated that their engagement with the network constituted a full-time job, as they constantly had to assist users with technical problems, explain configurations, or maintain equipment, expressed pride in what they had communally created. Anand, Gupta, and Appel's (2018, 26) remarks that infrastructures often excite affects and sentiment and produce a sense of belonging and accomplishment certainly apply to the case of SNET. However, I will now discuss how the strong emotions that the joint work on the network generated among its admins and users also resulted in rivalries and exclusions. These rivalries sometimes even led to serious conflicts over the organization and goals of the network, the use of limited technical and material resources, and how to deal with external pressures arising from the complex Cuban political context.

A constant source of dispute was, for example, the use of the funds generated by the obligatory membership fee that was introduced in 2015 to allow investments into infrastructure and repair. Each user had to pay 1 CUC per month that was shared between a fund at node level and a fund at pillar level. The introduction of this fee in many instances created conflicts between node admins and pillar admins as well as between node admins and users. Because the contribution was first introduced on the node level, it took long and tough negotiations between pillars and nodes to agree on how these funds should be split between the different levels of the network hierarchy (and, therefore, on who could decide how they would be invested). Users, in contrast, often complained about the lack of transparency about what happened with their money. There were nodes that were very open with their internal accounting and kept public lists that proved to the penny how the collected contributions were used. Other admins, however, began to see them as a monthly salary paid by users for being connected to them. Jaime, an informatics student from Habana Vieja, for example, recalled that the admin of his node used to spend the money as he pleased, and when something urgently had to be repaired, users frequently had to pay extra to regain connection. In some cases, this membership fee generated substantial revenues for admins who made investments into network technology to be able to connect as many users as possible (often several hundred) and have a steady stream of income.

Over time, the ongoing tensions around these funds brought to the fore differences in values, motivations, and managing styles and led to mistrust, unresolved hostilities, and unsettled scores. These tensions finally erupted when DaVinci, the admin of the RoG pillar, revealed that he and his developer team had created a virtual shop inside their modded World of Warcraft in which players could invest real money for virtual items (such as weapons or equipment) inside the game world. The shop functioned in a way that users could contact someone from the technical support and pay either personally in cash or by transferring credit to a mobile phone or national Internet account. In addition, they laid out plans for implementing a virtual currency. The earnings generated from this were meant to be invested in more servers to enhance the overall gaming experience. The shop was debated fiercely on TS and various SNET forums, and many members felt that this innovation de facto turned their beloved communitarian project into a business. Other pillar administrators felt threatened by DaVinci's plan because they expected that further investments into RoG's infrastructure would make the main game-hosting pillar even more popular, making it more attractive for nodes to connect to it and, eventually, tilting the delicate power balance between pillars even more in its favor.

To resolve the conflict, a meeting of SNET's main admins was organized in a café near the Malecón. Attendees described the atmosphere as tense and poisoned by the insurmountable differences that had built up between the leaders of various pillars. At the end of the night a vote was held in which four pillars backed RoG's advance and four, led by the pillar GNTK (pronounced "Genética"), voted against it, with the pillar Playa abstaining. The trench between both factions deepened when, shortly after this meeting, a subnode belonging to the Cerro Cerrado pillar asked to switch to Wifinet in hope of better connection speeds due to the geographical proximity. When the Wifinet pillar accepted the node (and thus also took over its collected membership fees), the fragile balance of power between the pillars finally broke down. Cerro Cerrado allied itself with the other pillars with which it voted against the online shop and blocked all IP addresses of the rival pillars. After two weeks of unsuccessful negotiations, the faction around RoG and Wifinet also disconnected from the other bloc. SNET had thus broken into two halves. From November 12, 2017, users who entered the network could no longer connect to services or people on the other side. Pedro, in his mid-twenties and a freelance software developer from Centro Habana, explained that many popular services, like the programmer forum Netlab, which had contributed immensely to SNET's technological development and the identification of its users with the network, disappeared soon after the split: "Once we fully realized that our members were left on different sides of the border, it didn't make sense to continue. The day that Netlab died, we all cried."

After the separation, several subnodes switched pillars because of advantages offered by the rival side. Many of the more idealistic users and admins (such as Pedro and Ernesto), for whom SNET was all about community and horizontal relationships and for whom Netlab was the central meeting point, frustratedly sold their network equipment and turned their backs on the project.[6] "Those who stayed in SNET were the gamers," Ernesto commented. For him, the spirit of collaboration that defined the SNET community had all but disappeared. SNET continued to exist in this partitioned form until summer 2019, when the Cuban government released new legislation on the operation of private Wi-Fi networks, forcing SNET makers to once again adjust to changing circumstances.

Modding Politics

While until summer 2019 it was illegal to import Wi-Fi equipment, the operation of a private network was not really prosecuted. Alvaro recalled that sometimes local authorities confiscated antennas, but this seemed to happen more because of a lack of understanding of what SNET was than because of an organized crackdown. To avoid shutting down their network, the SNET makers were therefore anxious not to come into conflict with the government. Like the casas matrices who compile el paquete, they proactively banned political debate and the distribution of pornography and further prohibited nodes connected to the Internet. Contrary to the idea of community networks as refuges of autonomy and strongholds against surveillance and repression (as, for example, elaborated in De Filippi and Tréguer 2015), SNET chose to actively self-censor members' freedom of expression in order to be able to survive in Cuba's authoritarian system. As a network primarily committed to gaming, its makers and users were not seeking any form of social change, but merely wanted to make up for the specific lack of connectivity they experienced, a motivation also connoted in the concept of resolver, which implies not radical reform or the challenging of power structures but rather the fixing of what is given with what is currently at hand. Being aware of the state's long-standing strategy of dealing with new cultural forms of expression and social movements by incorporating them into the institutional frames of revolutionary national culture, some pillar admins had early on sought informal meetings with state officials to improve SNET's standing with the government.[7] In particular, RoG's central administrator, DaVinci, had continued to explore ways to fully legalize the network. RoG developers, for example, frequently collaborated with the Joven Clubs de Computación by organizing workshops or making public presentations at these state institutions when they were releasing new World of Warcraft updates.

Government resolutions 98 and 99 from May 2019 changed the legal and political framework in which SNET was operating. With this new legislation, the Cuban state finally authorized the domestic use of Wi-Fi antennas (which many Cubans were already using to capture the signal from the public hotspots from their homes or their small private businesses like restaurants and vacation rentals, if these were located close enough to a Wi-Fi park). Yet, at the same time, authorities now regulated outdoor cabling, the use of the Wi-Fi frequency bands, and radio transmitter power, supposedly to reduce interference with the state-provided mobile phone network. And this was where SNET finally hit a brick wall, as the new rules restricted many of the frequencies the network relied on and limited transmission power to a degree that made the long-directional Wi-Fi connections between pillars impossible.

Reinier, an engineering student from Vedado, remembered how the day after the publication of the new laws, his whole node was on TS. After the administrators declared that with the new rules SNET would basically become illegal and opened the discussion channel, emotions ran high. A few days later, on June 1, the bloc composed of the RoG, Wifinet, Imperivm, and Habana-Este pillars published a communiqué on the official SNET Facebook page and on Twitter in which they asked the authorities to create a special license for SNET that would allow it to connect to the national intranet linking the network of the state-run Joven Clubs de Computación. They also asked their users to stay calm and not engage in the heated debates that had already flared up in the commentary section of the online state news outlet *Cubadebate*. In a second notification ten days later, they announced that they had already begun negotiating with representatives of the Ministry of Communication. With the future of SNET hanging in the air, many users were looking for ways to advocate for the survival of their network. Some of them staged a series of demonstrations inside different game worlds with the slogan "YO SOY SNET" (I am SNET) as their rallying cry. Images of these virtual gatherings were then posted on social media.

Yet, soon afterward the network split again over the question of how to react to the new regulations. While most users did not want to get into trouble with the government and the administrators on the pillar level tried to negotiate ways of integrating SNET into the state intranet, a small group of users decided to fight for the network's independence and autonomy. They took their disagreement with the new law to Twitter and later to the streets. Despite the explicit calls from pillar admins to refrain from public protests, these idealists organized a small demonstration in front of the Ministry of Communication that was attended by about a hundred people, something that rarely happens in Cuba. These online and offline protests were short-lived and quickly suppressed by state security (who visited and threatened their most vocal leader). The protests were also not backed by the silent majority of SNET users, who were too afraid or too politically apathetic and gladly willing to trade in their network's independence for an official and more smoothly running alternative. The manifestations did, however, generate considerable international media attention. Their hashtag, #YoSoySNET, was quickly picked up by dissident bloggers and activists and instrumentalized by Miami-based Cuban exile influencers such as Alexander Otaola for their anticommunist propaganda (see chapter 5). Hence, at the end of its existence, this network, born out of pure practical necessities and never encouraging political debate, ultimately inspired a moment of political action—which, ironically, was exactly what the government always feared and what led to the decision to ban it in the first place.

While the protest movement gave up shortly after the demonstration under pressure from the state (and its initiator publicly complained online that he felt abandoned by the pillar administrators), negotiations between pillar representatives and government officials continued. By mid-August, the pillar admins announced that they had reached an agreement with the Ministry of Communication and that SNET's nodes would be connected to Joven Club's network of 644 nationwide branch offices. For this, SNET's node operators would have to apply for a private network license and declare their equipment, while users would have to log in to the network with a registered account. This secured the survival of many of SNET's services (particularly its highly customized video games) but effectively marked the demise of its independence, anonymity, and decentralized framework. Due to the size of SNET and the slowness of Cuban bureaucracy, the process of merging the network with the so-called Tinored of the Joven Clubs was time-consuming and laborious. Gamers whose nodes were already linked to Tinored said they now enjoyed better connection speeds (most Joven Clubs are connected with fiber-optic cable) and could play over the network with users from all over the island. They now also had home access to all the other services within the national intranet, such as EcuRed (Cuba's politically sanitized version of Wikipedia), the blog platform Reflejos, the national email service, and Mochila, the official alternative to el paquete semanal. Some trademark SNET offerings, like particular games, forums, and TS servers, have already been migrated successfully. To facilitate this process, some of RoG's administrative and development teams were hired as Joven Club employees.

Profit-Oriented Networks

In December 2021, as the lockdown measures were gradually lifted and interprovincial travel resumed, Nestor and I had the opportunity to meet with the administrators of the gamer network in Camagüey. Our purpose was to understand how one of the smaller networks outside Havana was adapting to the new regulations. Marco, one of the founders, provided us with insights into the structure of CGNET (Camagüey Gamers Network), which, at its peak, connected approximately six hundred users. Unlike the network in Havana, CGNET is organized in a more centralized and hierarchical manner. The network revolves around two main administrators: Marco, residing in the La Vigia neighborhood in the north of the city center, and his friend Daniel, based in Montecarlo in the southwest. These two individuals are responsible for overseeing the network's two primary pillars, which includes twenty-three sub-administrators who connect users in different parts of Camagüey. As pioneers in establishing a computer network in the city, Marco and Daniel opted for a centralized and controlled structure right from the start, distinct from what they knew from SNET. They selectively accepted users who possessed the necessary equipment for a stable connection and carefully planned the infrastructure for each neighborhood where there was a sufficient user base.

To avoid potential issues with the authorities, Marco implemented a login portal that required users to enter a username and password when accessing the network. This precautionary measure ensured that no one could anonymously post any content against the government. Notably, Marco enforced this requirement even prior to the merger of his network with the official Tinored. CGNET users are charged 2 CUC for gaming access and 3 CUC for additional privileges, including multimedia content such as a vast server with movies and series. Marco's positive rapport with the authorities proved beneficial when Wi-Fi regulations changed. He successfully obtained all the necessary permits to continue operating his network and seamlessly integrated his existing structure into the local Joven Club network. Although he frequently lamented the bureaucratic processes involved (having to seek permission from the Joven Club for even minor configuration changes in his network), he remained committed to expanding its infrastructure.

As the state-run Tinored became more appealing for online gamers, offering the opportunity to interact with players from across the country, Marco decided to invest more in multimedia content to retain the attractiveness of his network. Amid the pandemic, he developed a streaming platform boasting over seventy terabytes of content. Users logging in with their credentials can access this platform from their tablets, phones, or smart TVs. When I logged in from my phone with a guest account, I was impressed by the platform's user-friendly interface, rivaling the smoothness of Netflix. Each movie had trailers and synopses, and upon clicking a film on the homepage, its soundtrack immediately began playing.

The state struggles to compete with the exceptional service Marco has created. His content surpasses the offerings of Mochila, the official alternative to el paquete, in terms of attractiveness. Additionally, Marco's platform is more affordable and accessible than the state streaming platform Picta, which consumes users' national data plans. Further, Marco proudly demonstrated for us the network's capability for video calls, and he generously provides each user with 8 gigabytes of cloud storage on his servers.

Although the new Wi-Fi regulations officially prohibit administrators from profiting from their networks, enforcing this rule proves challenging in practice. The authorities recognize that administrators need to collect funds from users to invest in maintaining and improving the network. The state itself does not contribute any resources toward expanding its Tinored network, which has experienced significant growth through the integration of local grassroots networks from all over the country. Instead, the equipment declared by net-work operators technically becomes state property, although administrators assured me that this has no tangible repercussions. Once the state ensured that networks like SNET and other grassroots initiatives no longer provided an anonymous and unsupervised communication space, it became more ac-cepting of the economic aspect of these networks. This shift in attitude allowed media entrepreneurs like Marco to sustain themselves through the user fees collected and devote their full attention to maintaining and expanding their services. Marco, in particular, can work full-time on his network, thanks to the income generated from user contributions.

Conclusion

With the integration of SNET, CGNET, and other grassroots networks into the state-provided intranet, the authorities once again successfully assimilated a private initiative into the institutional structures of the state. However, rec-ognizing the significant investment of labor, time, and materials required to build these networks, the state allowed their creators to maintain and even profit from them. This approach mirrored the arrangement with el paquete, wherein private media entrepreneurship was tolerated as long as it did not create autonomous spaces of communication and exchange beyond the gov-ernment's monitoring. Most network administrators across the country agreed to this arrangement, relinquishing their network's independence to avoid writ-ing off their equipment investments and losing a source of revenue. Yet even though the majority of users identified as apolitical and the administrators themselves did not actively promote societal change, the mere existence of networks like SNET still conveyed powerful ideas of successful independent self-organization. These ideas ultimately inspired a small fraction of users to voice their dissent against the government, as demonstrated by their protest in front of the Ministry of Communication. This protest foreshadowed sub-sequent demonstrations by intellectuals, artists, and journalists who rallied against increasing censorship a year later (see chapter 5).

The state's monopoly over media has been challenged not only by independent media entrepreneurs such as casas matrices or private network operators but also by the introduction of mobile Internet in Cuba, which coincided with the state's Wi-Fi law changes. Activists protesting the shut-down of SNET utilized platforms like Twitter and Facebook (and to a much lesser extent, the network itself) to publicly express their dissent, as these platforms have a broader national and international audience. With permanent Internet access now available to many Cubans, they are successfully leveraging global social media and messaging applications to establish new independent connections and communication structures beyond the control of the Cuban state. Today, dissent and political organizing primarily occur on major social media platforms, mainly controlled by U.S. tech giants, rather than within domestic networks that the government can easily shut down, as the case of SNET exemplified. As I discuss in the following chapters, expanded access to these services has shattered the government's control over how Cubans communicate with each other, creating new political challenges for the authorities.

Fragile Connections (2022)

Interactive installation consisting of three synchronized videos (16:49),
an offline web server, and DIY Wi-Fi antennas
BY NESTOR SIRÉ AND STEFFEN KÖHN

www.da.gd/snet

Starring: DctLove, Cassey, Kopek
Software development: Eduardo Pujol
Graphic design: Mauricio Vega
VFX: Helman Bejarano

Fragile Connections is an interactive installation cocreated in collaboration
with members of the SNET community and originally commissioned by the
2022 Warsaw Biennial. The installation's design is inspired by the antenna
towers erected by SNET admins on the rooftops of buildings to ensure uninterrupted Wi-Fi signal transmission. It mimics this tower structure, with
several discarded antennas, including homemade ones that we received from
administrators, attached to the top. The central part of the tower features three
screens, displaying a focus group interview conducted with three administrators. The interview captures the rise, fall, and subsequent rebirth of SNET,

FIGURE 3.2. Installation views of *Fragile Connections* from Biennale Warszawa, Warsaw (2022),
Osmo/za, Ljubljana (2023), and re:publica, Berlin (2023).

along with the personal experiences of our interviewees with the network. The conversation took place and was recorded in the voice channel of one of SNET's TeamSpeak servers. To establish a rapport among the interviewees, who were previously unfamiliar with each other, we invited them to engage in friendly competition within SNET's top three games: DotA 2, Battlefield 3, and World of Warcraft. These matches were screen-recorded and serve as the visual component of the three videos, captured directly on participants' personal computers.

The installation further reproduces the technological configuration of an SNET node, operating as a fully functional LAN. Audiences have the opportunity to connect to our network using their smartphones and explore various facets of SNET. Via a QR code, they can browse through a collection of modded video games and media content hosted on the network. They can also access SNET's code of conduct, which was collectively agreed upon by its members. Additionally, visitors have the opportunity to access and download research documents that delve into the history of SNET. These documents include several "screen walk" interviews Nestor and I conducted with users and administrators, offering insights into the network's evolution and functioning. The "screen walk" is an elicitation method that involves research participants providing comprehensive tours of the network on their personal devices, showcasing their activities while connected to SNET, the services they utilized, the games they played, and their interactions with other users. We captured these participant-led explorations using a simple screengrabbing software on their computers, later editing them into short videos resembling YouTube tutorials or gameplays—genres that our interlocutors frequently created or shared themselves (for a more detailed explanation of the method, see Köhn, 2025). The collection of research materials downloadable from our node further features photographs and videos that document the hardware infrastructure of the network. These visuals were either captured by Nestor or myself, or sourced from the private archives of research participants.

The work is complemented by a large infographic that spans an entire wall, illustrating the power dynamics and hierarchies within SNET. This infographic serves as a visual representation of our in-depth research into the network's decision-making processes. Furthermore, the installation includes a collection of stickers featuring the distinctive icons and symbols utilized by SNET members to express their digital identities on TeamSpeak. This interactive element provides visitors with the opportunity to playfully assume an SNET persona by affixing the stickers to their clothing, enabling them to self-identify as enthusiasts of specific football teams, games, or music genres.

The accompanying digital materials include the full videos of the focus group interview, installation views from the work's public exhibitions, three additional screen walk videos featuring SNET admins, documentation videos, and a PDF of the sticker set featuring symbols users employed to self-identify.

OMG!

Suavizante de Ropa
120 mn
Café Hola
60 mn

550   450   550   160

Picadillo
Mixto
Condimentado
115 mn

Perro caliente
120 mn

260mn   230mn

160mn   130mn   200mn

Aceite Premium sellado de 1 lt en
300 mn

QuickTime Player

10 % 8:40 p. m.

Medica
mentos
Gratis!!!

Creado por 453 [...] 1-7/5/20

Descripción

Este grupo fue creado con el
objetivo de compartir
medicamentos gratis. En este

Atrás   Info. del grupo

Intercambio y Trueque (Chat)
2,285 miembros

Vendo bolsas de Gel para el
cabello a tan solo 100 pesos
cada una.

Detergente líquido para lavar fregar
y limpiar con aroma de melon es
original

CHAPTER 4

Markets

On an unexpectedly muggy January afternoon in 2021, I found solace from the stifling heat outside in my friend Veronica's living room. Sinking into her slightly worn sofa, I couldn't help but appreciate once more the pristine interior of her apartment, most of it dating back to the 1950s, which she had inherited from her parents and barely altered over the years. As we were sipping the sweet and strong coffee she had prepared for us, our conversation drifted toward the practicalities of daily life. We had both run out of essential personal care products, which had transformed into coveted commodities as the pandemic dragged on, but we both were reluctant to endure the arduous wait outside state stores, only to find empty shelves. Determined to find shower gel for me and body lotion for her, Veronica turned to the digital realm, navigating the various messaging groups she had recently joined—an online network that had become an extension of the Cuban black market. "Take a look at this!" she exclaimed, her voice full of disbelief. She then handed me her slightly outdated but impeccably kept Huawei phone, displaying a photograph of a PET bottle, filled to about one tenth with a pink liquid. "People are selling portioned shampoo in *pepinos* [Cuban slang for big water bottles], and toothpaste wrapped in *jabitas de nailon* [disposable plastic bags], just so they can demand higher prices. That's how low we've sunk!"

In this chapter, I examine how the introduction of mobile Internet in Cuba, which eventually enabled permanent Internet connectivity, led to the creation of new online exchange networks outside of established government structures. I analyze how the inhabitants of Havana turned to semipublic group chats on messaging platforms such as WhatsApp and Telegram for

FIGURE 4.1. During the COVID-19 pandemic, semiprivate groups on social media platforms like Telegram became vital hubs where users traded scarce items, black market vendors sold hard-to-find commodities, and members formed support networks.

accessing goods that became increasingly unavailable during the pandemic. While the Cuban government (at least in the first months) quite successfully managed to contain the coronavirus, the lockdowns and the drying up of revenues from tourism strangled its already suffering economy. The state's lack of hard currency and the international border closures led to a significant drop in food imports, without which the government cannot feed its population.[1]

When the state-run supermarkets became empty, people in Havana and other cities began to use instant messaging applications to trade scarce items or exchange information about which stores still had products in stock. Such chat groups created digital spaces in which locals swapped goods such as canned food or medicine and informed each other about the availability of, for example, hygiene items in the official shops. Business-minded entrepreneurs began to set up home delivery services that they promoted via WhatsApp business accounts or Telegram channels, and black market vendors used these platforms to offer commodities unavailable everywhere else. Some of these groups quickly reached membership numbers in the tens of thousands and became an indispensable necessity for many Habaneros seeking to meet their basic daily needs. The widespread uptake of these global digital platforms in Cuba did not lead to a wholesale adoption of the colonizing capitalist logic that, according to platform studies scholars such as Nick Srnicek (2017), José van Dijck, Thomas Poell, and Martijn de Waal (2018), and Nick Couldry and Ulises A. Mejias (2019), is inscribed in them.[2] Rather, Cuban users appropriated them to manifest their unique perspectives on the world. What was truly captivating about the interactions taking place on these platforms is the intriguing coexistence of notions of solidarity and sharing, deeply ingrained in the population since the Revolution and perpetuated by the paternalistic socialist state, alongside the uneasy presence of market-mediated exchanges. This dynamic gave rise to ongoing negotiations of value and morality, as contested ideas were confronted and reconciled in the realm of everyday exchanges. As Veronica's comment about the creative packaging and inflated pricing of goods that are offered on the digital black market suggests, Cubans openly debated the shifting value of things and services during times of material scarcity. Within these messaging groups, discussions revolved around the market value of specific items sought for exchange in barter groups or the worth of services provided by black market resellers who charged elevated prices but facilitated easier access to goods by sparing their customers the hours-long wait in line at government stores, often even delivering them to people's homes. Such debates about the value of goods and services frequently expanded into discussions about the conflicting value systems that underlay interactions in these groups as exchanges based on sharing, nonmonetary redistribution, and ideals of solidarity coincided with opportunities for financial gain. The establishment of horizontal support networks through the widespread adoption of these platforms also paved the way for new channels of civic participation, consequently challenging the state's perception of itself as the sole provider of vital services.[3]

The Moral Economies of Online Exchange

To understand the ways in which Cuban people navigate the ambivalences of living in such a contradictory material and moral world, I examine their interactions within these digital market spaces through the lens of the concept of moral economy, initially formulated by E. P. Thompson (1971) and later taken up and extended into a general theory of popular resistance by James Scott (1985). The concept refers to how economic activities such as the production, exchange, and consumption of goods are underwritten by particular cultural sets of values. According to Thompson and Scott, these moral ideals emerge when political-economic models of state or market meet local value systems. While Thompson opposed using the term out of the context of the particular historical formation of hunger rioters clashing with emerging market forces he described, I follow Norbert Götz (2015) in attempting to expand the critical potential of the concept as an analytical tool capable of representing the workings of contemporary civil society. Götz proposes a dynamic understanding of moral economy as the outcome of an ongoing clash of norms that departs from Thompson's static notion of tradition. Such a perspective allows us to ask how moral economies are shaped not only from the top down but also from the bottom up, as a mechanism central to the economic choices of concrete actors and their ideational or material expectations of personal gain.

In the Cuban context, chat groups on WhatsApp and Telegram constitute important spaces where current shifts and ambiguities in the national moral economy can be observed. While the paternalistic socialist state never stopped publicly proclaiming the incommensurability of social values and market values (despite the fact that the structures of the corporate state were designed to extract profit from Cubans both within and outside the island), I argue that the day-to-day negotiations within these chat groups (the "bottom up") are much more complex than the ideological dichotomies in official discourse (the "top down") allow.

Cuban consumers have grown accustomed to navigating a complex and paradoxical food acquisition system, which consists of both subsidized markets administered by the socialist state and unsubsidized markets controlled by the corporate state. With their ration card—the famous *libreta*—all Cubans are entitled to a certain monthly quota of basic products, such as beans, rice, matches, chicken meat, and cooking fuels at heavily subsidized prices. Making such basics accessible to all Cubans was central to the welfare values of the Cuban Revolution, according to which necessities such as food are considered a human right, distinct from the world of commodities (Wilson 2013). Yet quantities have shrunk considerably over the years, and most Cubans nowadays find it impossible to survive on state provisions alone. Other products one might consider to be basic (such as cleaners and toothpaste) are available only in unsubsidized stores, for which Cubans nowadays are required to find access to hard currency. In addition, virtually all Cubans must rely on the informal or black market whenever the state markets fail to fulfill their needs.

On one hand, chat groups on messenger applications have come to extend older solidarity networks through which crucial information about the availability of goods circulates and people exchange or redistribute items others might need. On the other hand, they serve as digital extensions of the black market that has become a ubiquitous feature of the country's economic reality since the Special Period. Ideals of solidarity and disinterested altruism are key to the ideology of the Cuban Revolution (as propagated in the writings of the national poet José Martí or in Che Guevara's vision of the New Man) and many Cubans, while not necessarily committed to the political process, still identify with its underlying values (Kapcia 2008, 85). However, since the country's economic collapse after the fall of the Soviet Union and the economic and social reforms implemented by Raúl Castro since 2010, Cubans have increasingly been confronted with a different scheme of values. Scholarship on the economic transformations of the post-Soviet area has highlighted the moral crises that arose as the socialist moral order was superseded by the logics of the market and the official values of absolute egalitarianism gave way to informal relations of exchange that established their own moral norms (Ledeneva 1998; Mandel and Humphrey 2002; Patico 2008). While Cuba did not experience a similarly radical transition from one economic system to another, the neat ideological opposition between state socialist values of solidarity and reciprocity and the ethos of a market economy has begun to crumble as new economic opportunities give rise to new moral ambiguities (Andaya 2009; Holbraad 2017).

State-Run and Private Cuban E-commerce

During the initial coronavirus lockdown in Havana, the suspension of public transportation created disparities among households based on their financial status. Those who were more economically privileged, possessing cars, savings, and large freezers, were able to swiftly stock up on essential goods, resulting in empty shelves for others. The situation worsened for Cubans without access to hard currency when the state introduced the MLC stores, where most of the available goods were exclusively offered. These stores accepted only special bank cards loaded with foreign currency, either from abroad or through deposits in a Cuban bank. As a result, Cubans without foreign money found the government stores that accepted the local currency virtually empty. At the same time, the lines outside the MLC stores were unbearable, condemning even more affluent Cubans to endure hours of waiting in the scorching sun.[4]

During the lockdown, the state promoted online shopping through its delivery platform TuEnvio (Your Delivery), which had been operational since October 2019, in order to encourage consumers to stay at home. On this national website, purchases were required to be made using bank cards, and access to a smartphone and mobile data was necessary. As a result, this system, although not as exclusive as the MLC stores, effectively excluded those segments of the population that still relied heavily on the national food ration system. Additionally, TuEnvio faced significant challenges, including frequent congestion and downtime, as the system was unprepared for the high demand. This resulted in restrictions on order volumes and long delivery wait times. Products on TuEnvio often sold out within seconds, as thousands of users placed orders simultaneously, leading to frequent overselling.

Habaneros reacted to these ongoing problems via digital means. Quickly, WhatsApp and Telegram groups emerged in which people shared information about which shops were stocking which goods. Many of these groups specifically tried to dissect the inner workings of TuEnvio to help make the ordering process more transparent. In these digital spaces, people posted alerts when new goods became available through the official platform or informed each other about missing products in their shipment to let others know what might be out of stock. As the official system didn't state delivery times, users posted their sequential order numbers once they received their purchase so other customers could more or less calculate when it might be their turn to get their delivery. In this way, these groups also generated a space for socializing and community, as users frequently shared opinions about the quality and pricing of products, publicly complained about wait times, or simply shared the joy of finally receiving an order they had been expecting for a long time.

Given the dysfunctionality of the state-run platform, creative entrepreneurs were quick to set up their own home delivery services for everything from *croquetas* to homemade pies to handmade face masks and promoted these via WhatsApp. Many of these opportunity-seizing entrepreneurs were actually small-scale family businesses or individuals. By often focusing on just one product, they could minimize their economic investment and the frustrating chase after hard-to-find ingredients or materials. As the business model was based on home delivery, many families residing in less central or attractive neighborhoods took the chance to become active in the private market for the first time. Some of these entrepreneurs quickly professionalized, opening WhatsApp business accounts and presenting their products with appealing images.

Dairy, a thirty-one-year-old entrepreneur, offered vegetarian products such as yogurt, marmalade, pasta, bread, and vegetable boxes through her WhatsApp account. She delivered these products twice a week within the city center using her old Volkswagen Beetle. To ensure timely delivery, customers had to place their orders at least two days in advance. Prior to the pandemic, Dairy had worked as a mula, importing auto parts from Mexico and selling them on Revolico. However, when international air travel was suspended due to the lockdowns, she utilized her contacts with private food producers to start a new venture. As a loyal customer, I got to see Dairy regularly twice a week, barring any car breakdowns or fuel shortages. With all the workload falling solely on her shoulders, she faced enormous challenges. When she wasn't busy in her small kitchen preparing sauces, peanut butter, or pesto, she would be waiting in lines outside MLC stores to buy ingredients or driving to pick up bread from a private bakery or cheese and vegetables from a *finca* located an hour outside Havana.[5]

Taking the concept of online food delivery a step further, Marta Deus, a successful young entrepreneur, cofounded Mandao, the first app-based delivery platform on the island. Modeled after global players like Uber Eats and Grubhub, Mandao currently operates in nine cities across Cuba, offering meals from restaurants as well as food from farms and agricultural markets. With over a hundred drivers and around three hundred participating restaurants, Mandao thrived during the height of the pandemic when most restaurants were closed and delivery services were in high demand. When the Cuban government allowed the establishment of micro, small, and medium-sized enterprises (MSMEs, or MIPyMEs in Spanish) following the July 11, 2021, protests, Mandao became one of the first legally established MSMEs on the island, providing Marta's business with much-needed legal certainty after its rapid growth.

While Mandao primarily caters to Cubans living in Cuba, several online platforms abroad enable Cuban emigrants to send various items, including food, beer, and cell phone recharges, to their relatives on the island. Despite the U.S. embargo and Cuban laws prohibiting the sale of privately imported goods, companies like Miami-based Katapulk operate online marketplaces that deliver basic supplies, albeit often at staggering prices, to all Cuban provinces. For many Cubans living abroad, these platforms became the only means to support their families when the ban on international air travel disrupted the privately run mula business, which they previously used to send dollars or smaller items. The exact manner in which these online platforms circumvent legal restrictions in both Cuba and the United States remains unclear, but it appears that such businesses can exist only with ties to Cuban government officials. The authorities also seem to show less strictness toward private online companies, as they allow the state to receive much-needed foreign currencies through electronic payments, all of which are handled by state-owned Cuban banks.

The Digital Black Market

When the lockdown led to an increased shortage of basic goods, first WhatsApp and then Telegram quickly became the main platforms for black market sales in which products from TuEnvio and the government-run shops reappeared for higher prices. The term "black market" might sound overly dramatic in the Cuban context given its pervasiveness, overt visibility, and ordinary feel (Padrón Hernández 2012, 37). I still speak of a black (instead of an informal or unregulated) market to capture its moral ambiguousness and to mirror the emic term *mercado negro*. The black market had been part of Cuban social and economic life already during times of state-socialist normalcy, and even digital black markets have existed for a long time. Cubans came to rely on these alternative distribution infrastructures because private import businesses remained prohibited and the black market was the only outlet traders had.[6] Since 2007, the online classifieds website Revolico has facilitated the purchase and sale of foreign consumer products, such as smartphones, computers, and clothes by international brands, that are brought into the country by private importers. As these products are otherwise not available in government-run shops (or only with hefty tax markups), the impact of this site on Cuban everyday life cannot be overstated (Kirk 2015). Another popular "pre-widespread-Internet-access" sales platform was Google Groups, in which offers were circulated via mailing lists. Some years later, dedicated Facebook groups such as Timbirichi and

Compra y Venta en Toda Cuba began using social media to connect sellers and buyers. Over time, these various infrastructures became nested within each other, with one often serving as a platform for another, both as a copy and as an alternative, to reflect the varying degrees of users' access to the Internet. For example, Revolico is still distributed nationwide as an offline archive file in el paquete semanal but also operates a very popular Telegram channel, which, in turn, serves as a repository that links back to various local channels or Facebook groups that Revolico also maintains. This range of access options guarantees that the vast majority of Cubans can use the platform, whether they have permanent connectivity or not.

What changed with the pandemic and the scarcities it produced was that people began turning to digital platforms to purchase not only imported goods but everyday essentials. These items were previously bought from trusted contacts in people's neighborhoods when they were not available through official distribution channels, quickly transforming these platforms into infrastructures essential to people's daily lives. Studies of Cuban household consumption and food provisioning (Rosendahl 1997; Weinreb 2009; Pertierra 2011; Padrón Hernández 2012; Wilson 2013; Garth 2020; Mesa Cumbrera et al. 2020) have described how virtually all Cuban households after the end of the Soviet Union—and the ensuing economic crisis—have come to rely on the black market for the everyday struggle of putting food on the table. This body of scholarship has highlighted the ambiguous side-by-side nature of black and legal markets and the risks and moral complexities involved in navigating them.

Black markets essentially form a large unregulated gray area in which sellers operate without an official cuentapropista license and obviously don't pay taxes. With the issuance of these licenses (which exist only for certain professions), the Cuban state has cautiously created a regulated private sector since 1993 but still subjects economic exchanges to numerous restrictions.

Aside from privately imported goods (a venture that ground to a halt during the pandemic due to border closings and travel restrictions), the commodities that are offered in WhatsApp and Telegram groups are either resales at higher prices of things that *revendedores* (resellers) purchased in bulk when they were still available in the official shops or goods that were stolen from state workplaces.[7] While illegal, the latter is a practice that is widely considered licit (at least when done to support one's family and not for personal gains) and that many Cubans engage in because of the extremely low salaries paid in the state sector.[8] Given the high prices for consumer goods in the unsubsidized markets, it is downright impossible to subsist on a state salary alone. Taking things home from work and selling them on the black market or buying goods on the formal market and reselling them informally later at a higher price are therefore important strategies to diversify income, especially if one does not have a remittance-sending family member abroad or a more lucrative job in the private sector (which has also been hit hard by the lockdowns and the absence of tourists).

When people take things from their state workplaces, they prefer to speak of *la busqueda* (the search) or *la lucha* (the fight) instead of *robar* (to steal). They tend to see the moral breach as already committed by the state. In paying insufficient wages, the government has failed to fulfill its obligation to provide for its citizens (Padrón Hernández 2012, 125). Hence, what the state brands as illicit and the revolutionary ideology as immoral behavior might not necessarily be considered morally wrong by its citizens (Pertierra 2011, 130). Therefore, these infrapolitical strategies can be considered as a type of popular economic protest that effectively undermines the government's economic policies and contradicts the values publicly promoted by the state.

The moral evaluation of such practices, however, becomes more complicated when the goods that state employees "find" are taken more directly from customers. Hanna Garth (2020) vividly describes how many of her research participants were fully aware of how pilfering from the state—for example, when employees of a bodega stole from the subsidized food they sold to low-income households—effectively undermined the state's efforts to distribute basic goods to its citizenry. Navigating complex moral terrain, they attempted to undertake practices to meet individual needs that had the least negative impact on the collective good and tried to find ways to take from the state that did not come into conflict with their conscience.

The Cuban black market is characterized by a constant blurring of boundaries between formal and informal, legal and illegal, and moral and immoral activities. Ironically, the state silently tolerates most of these black market transactions, as vigorously prosecuting them would likely incite public outrage and protest. Instead, state media groups have chosen to scapegoat the *coleras*, women who endure long hours in lines (*colas*) in front of supermarkets to purchase and resell scarce goods, as the culprits responsible for the severe shortage of essential supplies during the pandemic (Bastian and Berry 2022). These women, often marginalized and economically vulnerable, were unfairly portrayed as immoral individuals using racist stereotypes, while the media narrative intentionally obscured the deep-seated inequalities of class, race, and gender resulting from successive waves of economic reforms.

The pervasiveness of the Cuban black market is further evidenced by the fact that the chat groups discussed in this chapter operate openly and are far from being secretive or "underground." They are publicly visible and accessible to everyone, and administrators do not limit membership nor perform background checks on new members. Nevertheless, buying on the black market has always involved some risks, as there is no formalization nor state control and the rights, obligations, and expectations have to be negotiated directly between buyer and seller. In fact, most Cubans have stories to tell about how they or some friends have been cheated in the black market: they paid in advance workmen who then never showed up or they received spoiled, contaminated, or altered foodstuffs. Some of these stories (often taking place during the Special Period) have even acquired the status of urban legends that serve as warnings of the dangers of putting too much trust in strangers.

In the face of these risks, buyers on the black market have always acted with great caution and employed several strategies in order to minimize the risk of being scammed or charged exaggerated prices. One central tactic is to buy only from acquaintances from their own social circle, thus relying on (as well as utilizing) the moral duties inherent in preestablished interpersonal relations. Creating and maintaining *confianza* (trust) is therefore a key prerequisite that undergirds most successful transactions on the black market. Anthropologists of postsocialist societies, such as Caroline Humphrey (2002), have shown that social infrastructures of trust become important whenever people cannot put trust in the state-managed system of distribution. Trusted sellers guarantee the quality of their goods with their reputation. Trust, therefore, constitutes an important form of social capital.

When things started to get scarce during the pandemic, people were quick to turn to digital media to widen their social networks to find access to rare goods. Virtually all Cubans I know are members of a plethora of WhatsApp groups that reflect their social circles (one for workplace colleagues, another one for old classmates from middle school, a third for neighbors, and so forth). Within these groups, people began to inform each other when someone knew someone who was selling crabs or "found" some cartons of eggs. People then organized collective purchases, either bringing cash to the trusted person who had contact with the seller or sending them money via Transfermovil, a new banking app that also allows digital payments between individuals.

The ethnographic literature on Cuban everyday economies describes the strongly gendered responsibilities in Cuban households where women bear a disproportionate share of the increasingly arduous, time-consuming, and unpaid domestic labor, encompassing tasks such as sourcing, preparing, and managing food (Wilson 2013; Härkönen 2016; Garth 2020). Consistent with female-attributed responsibilities of family care, nutrition, provision, and information exchange (in the form of "gossip"), I found that much more women than men were active in group chats dedicated to the selling or exchange of necessities like foodstuffs, hygiene products, or household appliances, and they also more frequently acted as providers of information. As Jennifer Cearns (2019) showed for the mula import system through which global goods reach Cuba, women are key to the success of these informal online circulation networks, and indeed the traditionally "female" sphere of information flow via gossip is of prime importance to their sustainability and expansion. In contrast, resellers—particularly those who also provided home delivery service—were more often men, which aligns with male-attributed activities such as acquiring things in the public domain.

The surge in the use of digital platforms for commerce has significantly altered the conventional operations of the informal market. On WhatsApp and Telegram, buyers now had to adapt their strategies for acquiring decent quality goods at a fair price to the realities of much larger and more anonymous online spaces. Instead of being able to build trusting relationships with sellers who have a vested interest in maintaining a good reputation, my respondents now had to find ways to exert some control in their interactions with unknown vendors in chat groups with hundreds, thousands, or sometimes even tens of thousands of members. Liritza, a woman in her mid-thirties from Centro Habana, exemplifies the cautious approach taken by individuals in the Cuban black market. She ensures her safety by taking screenshots of private chat conversations with sellers to preserve their phone numbers even if they might delete the messages. For her, participating in WhatsApp groups offers a sense of security, as WhatsApp openly displays the phone numbers of all group members. In contrast, she feels less secure in Telegram chats, where people can conceal their personal information. To enhance security during transactions, many individuals rely on applications like DITEL, an unofficial app that utilizes a leaked customer database of the Cuban telecommunications monopolist ETECSA. These applications enable users to associate a name and address with a specific phone number and vice versa. Participants in my research often mentioned that using such apps instills a greater sense of confidence as buyers can find out where the vendor lives should they have complaints later. These applications, which offer detailed information about any Cuban with a phone number (and deviate from Western notions of data privacy), enjoy widespread popularity throughout the country.

Research participants further based their decisions of whether to trust a particular seller based on the photos they post of their items. Maria, aged twenty-eight, for example, buys only products shown in their original packaging. Like my friend Veronica, she despised the many vendors who offered portions of personal care products in plastic cups so they could sell smaller units at higher prices (toothpaste became so scarce during the pandemic that black market prices peaked at the equivalent of eight dollars per tube). She felt that such practices give sellers far too many possibilities to manipulate either the quantity or the quality of the product. Héctor, a man in his fifties from Santos Suarez, was equally critical of traders who use images from the Internet instead of photographing the actual offered item, as he thinks this is usually done to hide minor damage in the hopes that a buyer who has taken the trouble to come to the vendor's house would still buy it, flaws and all.[9]

Another big concern for buyers in these message groups is overpricing, and the material value of products is constantly debated. Even though the evening news regularly reports on busts and seizures of illegally stockpiled products by the police, people know they depend on resellers. Buying from revendedores at least makes a product easily available without waiting in lines (or refreshing TuEnvio) for hours. Maritza, in her early thirties, for example, described how much she came to rely on resale groups: "My husband and I both work, I have a small child, I can't stand in lines all day, so the only option to eat is to buy from the reseller. . . . Many times, I have gotten milk for our child there, overpriced of course. But I have gotten it. In the stores it was impossible to find and yet in these groups, I have gotten it." Group members generally have a very clear sense of what they consider fair pricing and enforce this by publicly shaming vendors who charge exorbitant prices or even asking administrators to expel *abusadores* (abusers) from the group. In the reply section under the offers, people frequently debate if a price is "just." Adrismay, aged forty-two and from Playa, explained that "[this is] not like in capitalism, where you either accept a price or you don't. . . . There is a lot of social control in these groups." Rosa, a woman in her late fifties from Centro Habana, added, "People accept that you resell to earn one peso, two pesos, but people don't accept that you resell to earn five, or twice of what you paid for the item yourself." Thus, while Cubans do accept the capitalist values of individual profit maximization to some degree, entrepreneurial spirit is also confined by people's moral values, which are rooted in long-standing notions of community solidarity that have long shaped the national moral economy. Vendors are therefore expected not to put their pursuit of material gain entirely ahead of the need to offer goods to the public at fair prices.

Many resellers have reacted to this public policing by stating prices only in private messages to evade the scorn of the group's member base. One revendedor who sells coffee he receives from an uncle working at an MLC store told me that he doesn't want clients to come to his home. To avoid trouble with the police, he decided to only deliver to his customers. As Maria Padrón Hernández (2012, 91–94) notes, there is a certain complexity in Habaneros' attitudes toward overpricing and cheating in the black market. While people generally lament cheating and a lack of solidarity as big problems in contemporary

Cuban society, they also see street smarts, guile, and intelligent scheming as important skills for navigating the contemporary economic landscape. Where to draw the line between cleverly maximizing profit, profiteering, and immoral fraud is therefore subject to constant negotiations. While from the perspective of the state any resale is illegal and morally reprehensible, people's moral evaluation of individual vendors depends much more on whether the prices they charge remain within accepted limits.

Online Solidarity Networks

In addition to facilitating black market exchanges, messaging groups also became crucial hubs for "platform socialism" during the pandemic as people shared information, enabled the nonmonetary exchange of items, and fostered acts of altruism and solidarity.[10] In Cuba, information about what is available on both the formal and the informal market has long circulated informally within social networks rather than formally through the government-controlled press or media. In her ethnography of a small Cuban town in the 1980s, Mona Rosendahl (1997, 43) describes how the question of where one can buy something was the central topic of every neighborhood small talk. In present-day Havana, it is absolutely not uncommon to ask strangers who carry around plastic shopping bags where they found a particular item. Such sharing of information about the availability of things is a form of reciprocity in which virtually all Cubans participate. When the lockdown minimized the possibilities of encountering and chatting with people on the streets, the flow of such vital information seamlessly transferred to the digital realm, creating an online extension of this *boca a boca* (mouth-to-mouth) culture. In chat groups such as DondeHayEnLaHabana, Canal de Alertas TuEnvio, and Tuenvio_Reportes, people asked or informed each other about what government shops or the official online platform TuEnvio currently had on offer. On Telegram, some of these groups have elaborate bots (applications set to perform specific functions and interact with users) that allow users to better search the mass of accumulated information for specific products. Often, group members also add photos of the shelves or the waiting lines in front of the shops. In one case I heard of, such photographic evidence allowed a mother to buy diapers for her child even though the shop personnel had just cleared the shelves to resell them on the black market. With the help of a photo of the filled shelf that was posted in one of the groups a mere five minutes before she entered the store, she could pressure the shop assistants to sell her the diapers for the original retail price.

Other messaging groups such as Intercambio en La Habana, Cambios Utiles del Hogar, Kambalache, and Intercambio y Trueques dedicated themselves to the nonmonetary exchange of things. As most people buy all the heavily subsidized products they are entitled to with their libreta, and as articles on TuEnvio are mostly sold as assembled *combos* (packages), they frequently end up with more than they can actually consume of a particular product (for example, sugar or soybean oil) or stuff they don't actually need (because the detergent on TuEnvio was bundled with, for example, condensed milk). Therefore, they are happy to trade these things for something else. Barter is, of course, also a way to optimize one's resources and save money. When sixty-year-old Mónica, for example, was searching for shampoo, conditioner, and face cream during the height of the quarantine measures, she put a message on Kambalache in which she offered a set of champagne glasses in return. Many people also offer services in exchange for goods. Yusleydis, a woman in her mid-forties who had to close the beauty salon she co-owns with her husband during lockdown and was therefore without income, often traded facials or manicures for household products. The more distinct the products or services are that people seek to exchange, the more complicated these negotiations become, and they therefore often drag on for days of private message exchange.

The rules of most of these groups explicitly prohibit offers that involve money. However, my research participants' experience was that the monetary worth of the items or services to be traded remained a major reference point in the negotiations that ensued in private chat. When Mónica wasn't happy with the single shampoo package she was offered for her champagne glasses, she politely asked if there was something else the woman could propose to her. The woman sent along a photo of some canned meat that was available for trade, but Mónica was not interested. The woman did not have any eggs, which Mónica would gladly have accepted. However, they eventually agreed on a small sum of money in addition to the shampoo to compensate for the difference in monetary value. Teresa, aged twenty-seven, lamented that when she wrote users a private message about a certain product they offered, it turned out that they were willing only to sell it (not trade it), which was an increasingly frequent occurrence. She noted that in this way the lines between groups with differing objectives (and, therefore, also the lines between different systems of value) increasingly became blurred.

Resellers who offer their products only through private chat messages are difficult for group admins to control. Most of these exchange groups are intensely moderated, as their administrators do not want to get into conflict with the government. In the beginning, even groups that allowed monetary exchanges prohibited the sale of products that were also available in the official stores. This was an attempt to avoid the appearance of providing a platform to black market vendors, which meant that even when people wanted to sell things they received via the libreta and/or when they wanted to sell for the official price at the government shops, their posts were frequently deleted. This relaxed only when admins saw that the police did not interfere with what happened in these groups. However, rules that are still strictly enforced in all groups, regardless of their objective, include the prohibition of political debate and the dissemination of pornographic material, restrictions that are also observed by Cuba's private network operators and the matrices that compile el paquete semanal.

Moderators said that their work mainly consists of making sure that users stick to their group's organizing system to maintain its usability (for example, using the hashtags #tengo for things they offer and #quiero for things they are looking for). They invest considerable time and energy in deleting shady cryptocurrency offers and spam from Telegram bots. Moderating a group often resembles a full-time job. Some administrators silence the group and do not allow the publication of new information at a certain time in the evening until trades and sales start again at a fixed time the next morning. Several admins have developed business models that include charging fees for giving posts a wider visibility—for example, as pinned messages in Telegram—or providing users premium status that comes with special publication rights. Other groups are run by advertising agencies who see them as a form of publicity. In some cases, however, it is the users who impact (and sometimes even change) the objective of a group. The WhatsApp group La Farmacia, for example, initially was founded for the buying and selling of prescription drugs. However, an active user base ultimately changed the group's purpose to the free distribution of leftover medicines when some particularly active users took over the admin role from the founder of the group.

La Farmacia is only one of many groups that are purely driven by solidarity and altruism. Despite her membership in groups for the nonmonetary exchange of things, Mónica, for example, is particularly active in several messaging groups in which people donate medicaments. Because decades of scarcity have trained Cubans to be resourceful, its members recirculate medications left over from a cured illness or a deceased relative, knowing they may be desperately needed by others. As a trained pharmacist, Mónica has special access to various medicines and gives away whatever she can in these groups. She recounted, for example, responding to a message by someone looking for Nystatin, an antifungal medication. When an older man came by her house the next day to pick it up and asked what she wanted in return, she insisted that she didn't expect anything. The man disclosed to her that he was suffering from terminal cancer and needed the cream for treating the side effects of his therapy. Visibly moved, she showed me the message the man later posted in the group, thanking its members in general and Mónica in particular for their initiative. Teresa, who is also a member of this group, stated that she prefers to give unused medicine away instead of selling it because "as soon as you start selling medication you have lost your humanity. People take medicine because they need it, it's not like candy or soda!" Providing free health care to its population is one of the key services from which the Cuban state derives its legitimacy. Cubans therefore have come to see the public health sector as separate from the market.

The unselfish solidarity Mónica and Teresa pride themselves on is commonplace in Cuba and very much part of its citizens' DNA. Although many Cubans today might be tired of the state ideology, they nevertheless have accepted many of its values and principles into their own individual ethical convictions. Ongoing scarcity has taught Cubans that no one can survive on their own and that they must rely on others to get by. Showing solidarity and helping those in need therefore remains a defining component of contemporary *cubanidad* (Cuban identity), and these values are deeply inscribed in the national moral economy (Font and Quiroz 2005). Groups such as Mónica's create a space for people to live up to their moral values and craft their personal identity, while at the same time allowing them to experience themselves as part of something bigger. Mauri, the admin of a WhatsApp group linked to the Copincha makerspace he founded (see chapter 6), stressed that sharing for him also implies creation: of a community, a group spirit, or a new idea. This might explain why so many Cubans invest so much time and so many resources in documenting the inventories of supermarkets, crowdsourcing knowledge about the inner workings of the government e-commerce platform, or administering a messaging group to facilitate the donation or exchange of information or goods. All my research participants praised the friendly and supportive atmosphere prevalent in groups that are dedicated to nonmonetary exchange or donations. They all agreed on how these groups produced a kind of fraternity between their users and how they created communities around common interests. Indeed, more and more groups popped up throughout the pandemic that brought strangers together over the exchange of things beyond what was immediately needed for physical survival, such as plant seeds, vintage clothes, or books.

Online solidarity networks, which led to civil society initiatives, emerged even in areas that the government takes particular pride in for its organizational capabilities and focus on caring for those in need. When a devastating tornado struck Havana on January 27, 2019, citizens quickly mobilized to provide relief to those who had lost their homes. WhatsApp and Facebook groups were rapidly formed after the emergency, showcasing an unprecedented level of proactive response. Unlike in the past, Cubans did not wait for the state to take action. Particularly, more affluent individuals such as private-sector entrepreneurs, successful artists, and musicians took the lead in coordinating disaster relief efforts. While government workers focused on restoring essential services and initiating repairs in the most affected neighborhoods, private restaurants, as well as the renowned night club and cultural center Fabrica de Arte, transformed into temporary logistics centers. These locations collected donations of canned food, water, clothing, and blankets, which were then distributed to the tornado victims.

The state observed this grassroots commitment with a certain degree of suspicion as it threatened the state's monopoly on social assistance. For example, reggaeton stars Yomil and El Dani publicly complained on social media that a state security group hindered their food distribution efforts and expelled them from the disaster area. The unprecedented citizen mobilization even prompted President Miguel Díaz-Canel to acknowledge and praise the solidarity of the Cuban people on Twitter. While recognizing the people's acts of solidarity, he felt the need to emphasize that the planned economy and socialist government would always have resources so that no one is left helpless.[11]

This rapid and successful civil society self-organization held profound significance, especially for digitally literate young Cubans who actively participated in relief efforts. As I explore further in the next chapter, their tireless engagement in organizing donations, sharing information about collection points and transportation needs on Facebook groups such as Ayuda directa a La Habana Tornado (Direct aid to Havana tornado), showcased their ability to make a positive impact on the lives of their fellow citizens and harness digital technology to change Cuban society from the bottom up.

Conclusion

The WhatsApp and Telegram groups that facilitate access to food and essential goods for Cubans represent yet another example of a social infrastructure that fills the gaps left by state-run infrastructure projects. The influence of the state-orchestrated revolutionary process, which has deeply shaped the details of daily life in Cuba, has now been significantly disrupted by private initiatives that arise wherever the state cannot keep its historic promise to exclusively attend to all of the population's needs. In these emerging spaces, the revolutionary values promoted by the state no longer hold absolute sway. Instead, new moral economies are taking shape, and citizens are experiencing themselves as capable agents within civil society who can effectively organize and mobilize independently of state intervention.

The platformization of Cuba's informal economy further illustrates that the emergence of digital platforms does not automatically lead to the commodification of social relations and the colonization of human life, as scholars of "platform capitalism" (Srnicek 2017) assert. Rather, platforms often interact with economies at the peripheries or outside of global capitalism in complex and often unexpected ways, reshaping local practices while at the same time being redefined by them (Hobbis and Ketterer Hobbis 2022). WhatsApp, Telegram, and Facebook groups have markedly enhanced the visibility of products previously circulated only within personal networks, thereby significantly scaling up the black market and simplifying access for both buyers and sellers. Yet, at the same time, Cubans leveraged the expanded social connections afforded by these platforms to engage in nonmonetary exchanges, effectively recontextualizing these capitalist tools to reflect their communal ideals and practices and overriding the "moral economy of the platform" with the "moral economy of the users" (Bonini and Trere 2024). The ethnographic material discussed in this chapter reveals a contradictory reality in which users of digital platforms continually switch from engaging in networks of exchange that are marked by solidarity or even altruism to buying and selling (and sometimes even scamming each other) in uncontrolled and clandestine market spaces. Mónica, for example, can be a tough negotiator in groups that are dedicated to the barter of products, but she is so invested in her group that gives away free medications that she sometimes even spends money on medicaments she later donates to other members. In addition, many of my research participants entered the digital black market with the expectation of getting ripped off or cheated, but they still saw resellers as street-smart and accepted (within limits) that they make a profit.

Binaries such as profit versus solidarity or self-interest versus collective good—which are often used to describe the outcomes of Cuba's ongoing economic reform process—are therefore not helpful to explain the complex negotiations between morality and economy that users of these groups continually engage in as they shift between different value systems. As Maria Padrón Hernández (2012), Marisa Wilson (2013), and others have argued, these economic changes cannot be put down to something as simple as a new neoliberal ethos doing away with long-term revolutionary values or socialism slowly but inevitably giving way to capitalism. Instead, my examples show how noncommercial and commercial exchanges, the regular and the black market, legality and illegality, solidarity and personal gain all coexist and (often uneasily) overlap in these online spaces as people try to overcome scarcity and the devastating economic effects of the pandemic.

In her ethnography on food acquisition, Garth (2020) describes how her interlocutors rely both on networks of solidarity between family and friends *and* on illegal practices and channels (such as pilfering from the state and black market activities) in their daily pursuit of a decent meal. Hence, they engage in multiple affiliations that cut across different value schemes, some more communitarian, some more individualized. In a context in which no one can get by on their own, people rely on both interpersonal reciprocity and solidarity as well as market-mediated exchanges to survive.

The lack of government rules and regulations requires users and admins of the chat groups to jointly negotiate agreements on fair pricing, a particular group's objectives, or ways of building trust among members. Thus, people's moral behavior in these digital spaces has much more to do with what they feel is right and wrong than with dichotomies between socialism and capitalism. Cuba's present-day moral economy is predominantly driven by grassroots efforts and the everyday moral negotiations of its citizens. Rather than being a top-down economy dictated by the egalitarian value system promoted by the socialist state or the market-driven logic of digital platforms, it is a bottom-up economy that emerges from the collective actions and choices of individuals.

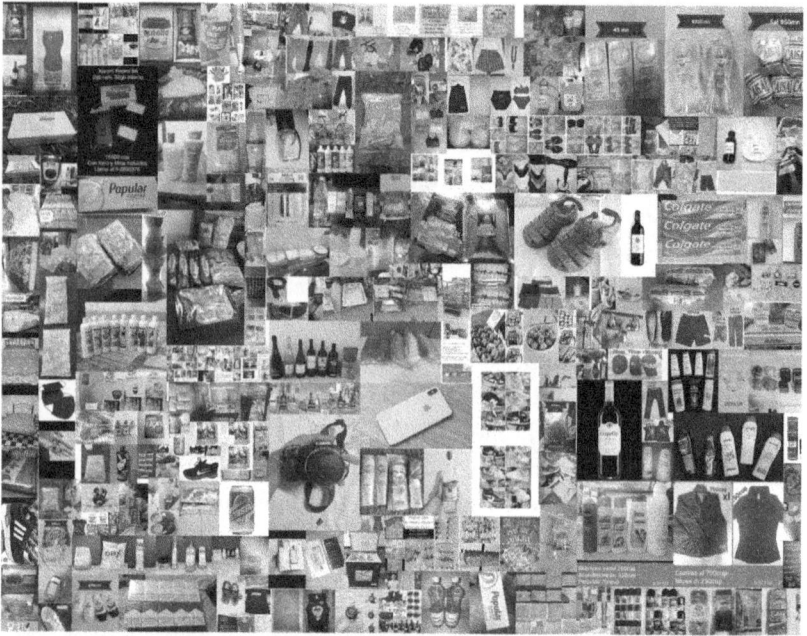

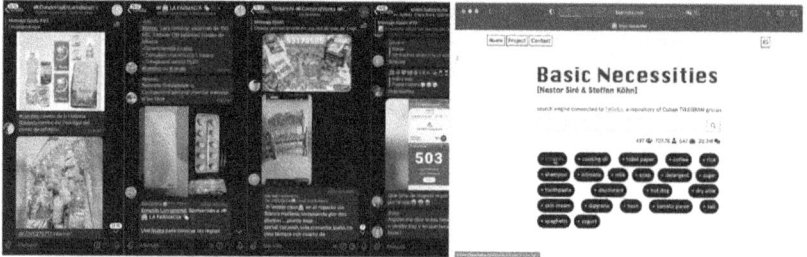

Basic Necessities
[Nestor Siré & Steffen Köhn]

Media Wall

Basic Necessities (2021)

Site-specific video installation and online database
BY STEFFEN KÖHN AND NESTOR SIRÉ

www.da.gd/markets

Database development: Rafael Rodríguez

Basic Necessities is a site-specific video installation and online metasearch engine commissioned by the Photographers' Gallery London. With this project, Nestor and I aimed to visually document the functioning and aesthetics of the Cuban digital black market. We showcased real-time interactions from four prominent Telegram groups on the gallery's media wall through a livestream. These groups included two major platforms for black market sales, a group focused on sharing information and advice about the state-run distribution platform TuEnvio, and a group facilitating the exchange of leftover medicines. Together, these groups boasted over 300,000 members. For one month, gallery visitors had the opportunity to observe the fluctuating demand for products on the island, the availability of goods through state distribution channels, and the scarcity of certain items. They could also track the development of prices of essential food items like coffee or chicken meat. By presenting a live documentation of Havana's current economic

FIGURE 4.2. Installation views of *Basic Necessities* from the Photographers' Gallery London (2021) and screenshots from the project's online extension.

situation, the installation revealed the fascinating social dynamics at play within these meticulously organized groups, shedding light on the underlying social currents.

In particular, the work displayed the eclectic and creative uses of product photography in these contexts. The images posted by sellers often encompass a spontaneous blend of appropriated stock photography and casual snapshots that provided glimpses into the daily lives of Havana residents. The aesthetics and usage of photography in this context are shaped by the specific material limitations and characteristics of the Cuban sociotechnical environment and the platforms on which they circulate. To save data, most users configure their phones so that images on Telegram or WhatsApp are visible and downloaded only when clicked on. Consequently, potential buyers first read the offer before deciding whether to view the accompanying photos. As a result, product photos must be impactful, as sellers rarely capture their products from multiple angles like on platforms such as eBay or Craigslist. Therefore, sellers often create collages with multiple products in one image and use simple cell phone applications to add text, such as prices or additional descriptions, directly onto the image. While this creates a scattered aesthetic, it facilitates easier sharing of the offer across multiple groups. Sellers frequently include their phone numbers directly on the image to ensure that their contact information remains intact even when their offer is shared by others. The crucial aspect of these product photos is that they form the basis on which buyers decide whether to trust a particular seller. Photography is therefore deeply intertwined with the social dynamics of these chat groups, serving not only for promotion but also for providing reassurance in informal market transactions.

Due to the fragmented nature of the digital market, where multiple competing groups exist, locating specific products and services can pose a challenge. To address this, we developed BasiCuba, an online extension of our project. BasiCuba functions as a metasearch engine, accessible through the Unthinking Photography web platform of the Photographers' Gallery. For the duration of the exhibition, it allowed users to conduct live searches across hundreds of Telegram groups. It draws on a database that was compiled by software developer Rafael Rodríguez, who has extensive experience in developing bots and tools for Cuban businesses on Telegram. During the exhibition, bots were installed in 476 groups (with a total user count of approximately 707,000), resulting in a database that includes approximately 17.8 million saved posts. Even after the conclusion of the exhibition, this database remains accessible, documenting the searches conducted by users throughout the exhibition period. BasiCuba thus serves as a comprehensive record of the phenomenon of this digital marketplace.

The digital supplement to this interlude features video documentation of the project, exhibition views, a link and documentation to the metasearch engine, and a collection of visual research materials that capture the distinct aesthetics of product photography within these groups.

PRIMERA QUEDADA DE YOUTUBERS EN CUBA

⊙ ⊘ First YouTuber reunion on Calle G in Havana.JPEG

⊙ ⊘ Pedrito Peluketa V154 Octubre Reupload ...

¿Están DE ACUERDO los cubanos con su SISTEMA?! ¡SON FELICES?!!🤔😳

⊙ ⊘ Demonstration in favor of the Animal Protection Law...

CUBA
contra el
MALTRATO
ANIMAL

⊙ ⊘ Cuban Youtubers documenting t...

ADIOS
CUBA!

⊙ ⊘ Cuban Youtubers documenting from emigration to.PDF

Nicaragua

Cuba

30:56

CHAPTER 5

Publics

On September 8, 2018, the Cuban government conducted a test of its soon-to-be-launched 3G mobile Internet service, offering a free 50 MB data package to prepaid customers using Internet-enabled phones. I had spent that day with a clique of young YouTubers around Dina and Adriano, known online as DinaStars and ComePizza, whom I had recently befriended. At that time, their friend Jhans had access to a spacious Airbnb in Habana Vieja, courtesy of his Russian partner who had extended the rental period beyond his stay in the country. So Jhans invited all his friends over, and they spent almost a whole week drinking cheap rum, eating pizza, and producing new content for their YouTube channels. Whenever they left the apartment, it was primarily to visit a nearby Wi-Fi hotspot, where they diligently uploaded their latest videos. Now that they suddenly had Internet access on their phones, not a single one of them could resist the allure of the novel possibilities at their fingertips. As I observed Dina, Jhans, Adriano, and the others instantly immersing themselves in live streaming on Facebook, sharing Instagram stories, and wholeheartedly engaging with the various social media features that thrive only with a consistent mobile Internet connection, it became unmistakably clear to me that this newfound ability to capture and instantly share every aspect of their lives would have a profound impact on the fabric of daily existence in Cuba.

In what follows, my focus is on exploring how young Cubans, the generation Ted Henken (2017) has called the island's "digital millennials," utilized their increasing access to the Internet to seize new economic opportunities

FIGURE 5.1. With the advent of mobile Internet, young social media personalities quickly made their mark on the Cuban public sphere. The rise in social media access sparked new civil society movements, including animal rights advocates, and fueled protests like the artists' demonstrations against increasing censorship in front of the Ministry of Culture. When the protest movement subsided, many of these influencers left the country, turning their video documentation of emigration into its own distinct genre.

within the emerging private sector, established transnational networks, and advocated for greater political liberties from the government. This generation, born approximately from the mid-1980s to around 2000, has experienced the economic turmoil of the Special Period during their childhood and witnessed Raúl Castro's economic reforms as teenagers or young adults. I developed a particular interest in this segment of Cuban society due to their enthusiastic adoption of digital technologies. Unlike their parents, many of them felt less connected to the socialist and solidarity ideals of the Revolution, yet they still aspired to contribute positively to their society—albeit on their own terms. As tech-savvy individuals, including digital entrepreneurs, influencers, artists, musicians, and independent online journalists, they actively engaged with social media platforms to generate income, established new connections (often on an international scale), built personal brands, advocated for social, political, or economic reforms, and found their voices within Cuba's gradually diversifying public sphere. By harnessing the power of digital technology, they experimented with novel forms of social organization and explored new avenues for cultural and political expression.

My immediate fascination with this generation of young Cubans led me to create the ethnographic documentary *Dinamita* (2022), which is part of the digital supplement of this book and will be discussed in more detail in the next interlude. In many ways, the film serves as a companion piece to the narrative presented in this chapter. Over the course of several years, I followed Dina, Adriano, and their group of friends, capturing their journeys and experiences. When we began filming in 2018, I had no idea where their obsession with the video-sharing platform would lead them. While the documentary primarily focuses on the personal narratives of Dina and Adriano, blending our observational footage with the videos they created and shared themselves, I aim to provide a broader account in this chapter of how the young Cubans I encountered during my research have made the Internet and social media their own.

It is important to note that the young adults whose lives and lifestyles are described here do not represent the average Cuban youth. They belong to an avant-garde group, even within Havana, where the private sector and digital infrastructure are more developed compared to the rest of the country. However, I argue that they have a significant cultural impact as they bring new ideas and demands into the political realm. Some of them have built substantial audiences, often composed of peers or even younger individuals, who listen to them and engage with their content. Although many of these youngsters may refrain from making overtly political statements, they still promote new forms of material and cultural consumption, driving what Edwin J. Nijssen and Susan P. Douglas (2011) refer to as "consumer world-mindedness." They increasingly live their online lives as global citizens, absorbing international cultural trends and Western models of consumption.

The growing Internet access among young Cubans, as explored by Cuban sociologist Yoannia Pulgarón Garzón (2021) in her study on contemporary youth identities, has fostered this adoption of global cultural influences. One notable example of the new social-media-driven teenage subcultures she examined was the Durakitos, a community that emerged in 2018, coinciding with the launch of the 3G mobile network, but began to fade as quickly as it appeared with the onset of the pandemic. The Durakitos embraced an eclectic mix of elements from reggaeton music videos, Latin trap, and Korean K-pop. They adorned themselves with branded clothing and, even prior to the pandemic, decorated masks. They carried extravagant accessories like stuffed animals and organized elaborate photo competitions in which they divided themselves into teams. They ventured into peculiar locations (one team that took selfies in a cemetery caused significant outrage in the official media) and shared their photos on dedicated Facebook sites like Vota Vota, where they voted on who was cooler or more *duro* (a reggaeton term that translates as tough or sexy). To accumulate as many likes as possible, they developed a myriad of strategies, such as imitating famous artists they identified with in their photos. Given that all of these activities required constant Internet access, the leaders of these teams took on the responsibility of ensuring that all members could be online during their gatherings, either by purchasing enough top-up cards or by sharing a mobile hotspot with their phones (Pañellas Álvarez 2021).

When I returned to Cuba approximately a year after the launch of the 3G mobile network, I was amazed by the profound impact that mobile Internet and constant connectivity had already made on the daily lives of the young Cubans I had become friends with. They now navigated the city using Bajanda, an Uber-style ride-hailing app, and ordered food through Mandao, the Cuban equivalent of the Grubhub delivery platform. Dina, Adriano, and their friends, who used to distribute their videos through the paquete semanal or upload them late at night at the Parque Trillo hotspot for better connection speeds, now regularly shared story videos on Instagram. They became recognizable figures in their communities, started speaking out on social issues more frequently, and began to garner increasing attention.

In the following sections, I look at how Havana's digital millennials navigated the intricate economic, sociopolitical, and technological transformations happening around them, actively contributing to these changes in various ways. I explore how members of this generation have become active participants in significant recent political events that have greatly influenced the relationship between citizens and the state on the island. These events include the new economic opportunities following the historic visit of President Obama, the rise of independent online journalism within Cuba, the protests by young artists and intellectuals against growing censorship, and, last, the unprecedented public demonstrations that took place on July 11, 2021.

"17D," American Tourism, and the Airbnbification of Havana

Many of my research participants began exploring digital entrepreneurship during the brief period of increased U.S. tourism that followed the events of "17D," a term used by Cubans to refer to December 17, 2014. On this significant day, Barack Obama and Raúl Castro jointly announced their intention to restore diplomatic relations between the United States and Cuba after more than five decades of suspension. Obama's subsequent visit to the island in March 2016 brought about several exceptions to the U.S. trade embargo, which had a profound impact on the economic prospects of young Habaneros. One key development was the relaxation of travel restrictions to Cuba, enabling U.S. cruise ships to visit the Cuban capital and allowing the online lodging service Airbnb to operate on the island. These changes created new opportunities in the rapidly expanding tourism sector. David, in his late twenties, who had established himself as an informal tour guide, elaborated on this point: "With the cruise ships, you suddenly had thousands of people streaming into Havana on any given day. And the cruise ship tourists were not like regular tourists. The cruise ship tourists were only in Cuba for twenty-four hours, so they wanted to see and experience everything. If they had a thousand dollars in their pockets, they would spend every penny." Many young Cubans, like David, who spoke some English and had university degrees, were not interested in taking up poorly paid state jobs. Instead, they sought ways to benefit from the surge in urban tourism. They quickly discovered that Airbnb provided a digital platform that seamlessly connected them with foreigners. Airbnb not only facilitates accommodations but also offers cultural activities known as "Airbnb Experiences." The entry of Airbnb into the Cuban market was initially successful. The company leveraged the existing network of private accommodations known as *casas particulares*, which had been established in 1997 when the government allowed citizens to rent out rooms in their homes to tourists for additional income. However, prior to Airbnb, it was extremely challenging to book these rentals from abroad, and tourists had to find them upon arrival by asking taxi drivers or restaurant staff. When Airbnb launched its business in Cuba, approximately four thousand out of an estimated twenty thousand casas particulares immediately joined the platform, making it Airbnb's fastest-growing market in 2015 and 2016. While attractive housing in tourist areas was often concentrated with prerevolutionary and revolutionary elites, who could now reestablish themselves at the top of the emerging socioeconomic hierarchy, the marketplace for cultural experiences on Airbnb required little more than creativity, intercultural competence, and an understanding of what foreign travelers might be interested in and willing to pay for. As a result, many of my research participants utilized their cultural capital and social networks to offer guided tours of the city's subcultural nightlife, explorations of queer or Afro-Cuban Havana, street photography classes, or yoga sessions in El Bosque, the capital's green space. For example, my friend Yasser organized a successful "Havana Underground" bike tour. During this tour, he explained the city's street art and tattoo culture, introduced his clients to Cuba's informal economy, and took them to a paquete copy store before ending the day at a locals-only bar.

Yasser's journey reflects the inclusive and high-quality public education system in postrevolutionary Cuba. Being born into an Afro-Cuban family in the remote town of Nueva Gerona on Isla de la Juventud, a pine-covered island three hours away from the mainland by boat, he successfully passed the entry exams to attend the capital's most prestigious institutions. He first went to La Lenin, a renowned boarding school that many children of the political leadership attended, and later pursued software engineering at the prestigious University of Information Sciences, which was established by Fidel Castro to develop a national software industry. Upon graduation, however, Yasser, like many other alumni of La Lenin who chose to migrate or work outside the state sector (Berg 2015), decided not to join the ranks of Cuba's socialist elite. Instead, he participated in Cuba Emprende, a business training and advisory project led by the Catholic Church of Cuba. Through this program, aspiring entrepreneurs received classroom teaching on various aspects of business, such as marketing, management, finance, accounting, customer service, and sales. Yasser utilized this knowledge and expanded beyond Airbnb. His bike tours had become incredibly successful, allowing him to employ five people. After some earlier unsuccessful attempts at running businesses with partners, including opening a hostel for backpackers, he obtained a cuentapropista license as a "lessor of leisure equipment" and established his own company called Citykleta. Citykleta, which Yasser describes as a "social-impact business," offers guided bicycle tours and events for tourists while promoting the bicycle as an alternative and environmentally friendly mode of transportation. Yasser also advocates for the development of bicycle-oriented infrastructure and the implementation of supportive policies.

Like many of the digital millennials I encountered, Yasser had entrepreneurial aspirations that were driven not solely by financial success but also by a desire to positively influence Cuban society, which is why he participated in various forms of online and offline activism. During his tours with foreigners, he acted as a cultural translator, offering insights into the intricacies of contemporary Cuban reality. He eagerly absorbed knowledge about emerging global cultural trends and advancements from his clients. Initially, Yasser found it challenging to adapt to Airbnb's platform capitalism model and assign a monetary value to something he used to do for free. In 2012, when Internet access was limited to government workplaces, universities, tourist hotels, and privately run illegal casas de conexión, he had been one of the first Cubans to open a couch-surfing account and invite backpackers to stay at his small apartment in Vedado without charge. Unable to travel himself at the time due to travel restrictions imposed by the Cuban government (which were relaxed in 2013), hosting couch-surfing guests became his way of exploring the world. Although it was illegal for foreigners to stay in Cuban homes without immigration permission and payment, Yasser took the risk. When a woman he met in Havana invited him to visit her in Berlin, he became fascinated by the bicycle activism he encountered in the German capital. He brought back the concept of "critical mass" events, where cyclists gather at a specific time and location to ride together through their neighborhoods, reclaiming the streets and protesting a transport policy that prioritizes motor traffic. Since 2015, Yasser has regularly announced critical mass bike rides in Havana through social media. Initially, only a few foreigners based in Cuba participated, but over time his events attracted a diverse mix of local cycling enthusiasts.

Yasser remembered the Obama years as a time of great optimism, anticipation, and enthusiasm, when everything seemed possible. Havana suddenly felt like the center of the world. The city emerged as a magnet for tourists, eager to experience it before the anticipated transformation ushered in by American capitalism. Lifestyle journalists wrote cover stories about the local arts and music scenes, and iconic acts like Major Lazer and the Rolling Stones held free open-air concerts, each attended by hundreds of thousands of people. The Hollywood action movie franchise *The Fast and the Furious* even filmed one of its car chasing scenes on the Malecón. Yasser himself had the opportunity to socialize with the CEO of Airbnb at one of the newly opened rooftop bars in Centro Habana, and Google's head of Cuba joined him on his bike tours before interviewing him for his podcast.

During that time, a significant amount of money flowed into the city. As the Obama administration eased restrictions on remittances, Cuban Americans began sending larger sums of money to their relatives on the island. These funds not only were used for day-to-day expenses but often also served as start-up capital for small businesses. Many Cuban exiles returned to Cuba, taking advantage of the 2013 reform of the migration law, to invest in the real estate market or apply their resources and knowledge of global cultural trends to establish small companies. According to the National Statistics and Information Office (ONEI), in 2016 alone 14,000 individuals decided to repatriate, and in 2017 that number was 11,176 (Jiménez Enoa 2018). Among them were some of the most successful entrepreneurs today, such as Marta Deus, the founder of the Mandao food delivery service and the business magazine *Negolution*. Repatriated entrepreneurs also pioneered the establishment of consulting and marketing agencies on the island and actively advocated for greater economic freedoms from the government.

Business-savvy young Cubans also had their share of the profits from this boom as adventurous tourists willingly paid between thirty and fifty dollars for an Airbnb experience. Many of my research participants were able to earn monthly incomes ranging from two hundred to a thousand dollars through the company. Airbnb's platform capitalist model, however, did not work as frictionless on the island. The money Cubans earned on the platform thus often found its way into their pockets through a unique channel. Given that many of them lacked international bank accounts and U.S. banks were cautious about entering the Cuban market due to concerns of violating U.S. sanctions, Airbnb partnered with VaCuba, a Miami-based remittance service, to facilitate cash payments. VaCuba operated a network of intermediaries who personally delivered payouts in the local currency directly to the homes of hosts and activity providers. This alternative infrastructure provided a workaround for the restrictions caused by Cuba's isolation from global payment traffic (Tankha 2021).[1]

The optimism of the Obama years began to fade when hardliners in the Cuban government started to put brakes on the opening process they feared was going too far, and it abruptly ended when, in 2017, the newly elected Trump administration began to roll back Obama's engagement policies. It de facto eliminated individual travel because the permissible category of people-to-people educational travel under which most Americans had come to Cuba now required an official organization that sponsored such trips. Cruises, one of the most popular ways for Americans to travel to Cuba, were once again banned from calling the island's ports. Remittances were limited to one thousand dollars per person per quarter, and from October 2020 the Department of the Treasury prohibited the processing of remittances through any entities on the Cuba Restricted List, which includes various institutions of the corporate state, including both Cuba's largest commercial conglomerate Fincimex and its owner, the military holding company GAESA. This caused Western Union to close all of its 407 nationwide branches and suspend operations in Cuba. In January 2021, in its last days, the Trump administration even redesignated Cuba as a state sponsor of terrorism. This decision also negatively impacted EU tourism to Cuba as Europeans traveling to such designated countries see their ESTA (Electronic System for Travel Authorization—a visa waiver for short U.S. visits) eligibility revoked.

These new U.S. policies severely limited the influx of travelers, and the coronavirus pandemic further devastated the tourism industry. Consequently, Yasser and many other young entrepreneurs in that sector decided to put their businesses on hold and stop paying for their cuentapropista license. Yasser sought alternative funding for his projects and ideas, eventually receiving support from European embassies and international NGOs. Similar to other Cuban digital millennials, he became discouraged in his attempts to earn a living from the tourism industry due to these external circumstances, but also because of the lack of improvements in state regulations governing private businesses. When the government announced its plans to open up the private sector in February 2021, it removed the list of 127 permitted economic activities that the self-employed had utilized with great creativity (my interlocutors who offered Airbnb experiences, for example, did so under licenses such as photographer, event organizer, or contract worker). The new regulations now introduced a negative list, allowing private entrepreneurship in most areas except for a few specific professions that remained the monopoly of the state. Unfortunately, many young Cubans were dismayed to find that professions they had hoped would be legalized, such as tour guide or architect, were included on the negative list.

David, on the brink of launching his own travel agency, drawing upon the contacts and expertise he had acquired through designing and coordinating four immensely prosperous Airbnb experiences, voiced his frustration with the government's perception of individuals like him as competitors. He bemoaned the lack of understanding within the state's inflexible structures, which prevented state-run tourism agencies from providing the same level of personalized tours or activities that he had meticulously crafted. Consequently, he decided to explore the option of registering his agency outside of Cuba, enlisting the assistance of a Cuban partner residing in Panama, in order to legally operate on the island. While recognizing the profitability of the tourism sector, particularly in light of widespread inflation that made foreign currency payments highly appealing, David harbored no illusions that tourism would swiftly rebound to its pre-Trump levels.

Building a Personal Brand

Increasing Internet access had a direct economic impact on my research participants, enabling those who were highly active on YouTube and social media to build personal brands and monetize their online presence. Dina, Adriano, and their group of friends all began creating videos for enjoyment, without considering any financial incentives. In fact, they even shelled out money to include them in the paquete or to upload them in the Wi-Fi parks. Over time, they discovered their own unique content and developed their individual voices, initially drawing inspiration from Chilean, Peruvian, and Cuban American YouTubers they found in the *humor* and *interesantes* folders of the paquete, such as HolaSoyGerman, Mox, or Los Pichy Boys. Dina, for instance, initially started her channel by providing fashion and makeup tips, influenced by the Colombian beauty vlogger Pautips (although she adjusted them to the Cuban context, considering the limited availability of beauty products). However, upon realizing that many other female YouTubers were doing the same, she began to explore other genres such as humor or YouTube challenges. Even during this period when content creation was primarily seen as a hobby among young Cubans, some of my research participants were already exploring avenues to generate income from their material, particularly those who discovered they had an international following. For instance, Dina and Adriano were able to earn approximately three hundred dollars per month by creating humorous video content for the Miami-based web platform Cuba Cute, which specifically targeted a Cuban American audience and actively sought out content creators from the island.

Before the advent of mobile Internet in Cuba, Cuban vloggers heavily relied on offline distribution methods such as the paquete semanal and the peer-to-peer file-sharing app Zapya to circulate their videos. In order to establish a feedback channel with their local audience, who were hesitant to use their limited Internet access to go on YouTube's comment section, many video bloggers would include their phone numbers at the end of their videos, enabling followers to contact them via text messages. The introduction of mobile data and improving connection speeds then significantly streamlined the distribution of their content and communication with followers, relegating the challenges of Cuba's patchwork digital infrastructure from a prominent to a secondary concern. When young Cubans finally gained the ability to utilize

social media platforms as originally designed—without the need for improvised or alternative channels for dissemination or communication—this infrastructure, in the sense described by Susan Leigh Star (1999), progressively
receded into the background, becoming less visible due to its smooth and
frictionless operation. However, their increasingly seamless access to digital
platforms was still constrained by the U.S. embargo, which barred American
companies from providing some of their services, particularly their financial
ones, on the island. For example, Cubans residing in Cuba are unable to monetize their channels directly on YouTube, a feature normally available after
reaching certain milestones like one thousand followers and four thousand
playback hours within the previous twelve months. To receive a share of the
platform's advertising revenue, my interlocutors therefore had to register their
accounts under the name of a trusted individual living outside the country,
subsequently relying on yet another human infrastructure that circumvented
Cuba's economic isolation. In Dina's case, she registered her account under the
name of her aunt, who lived in the United States and then sent her YouTube's
monthly payments through one of the informal money transfer networks that
connect the countries via the mulas who carry cash and goods back and forth
between relatives on the island and the diaspora. To collect the money, Dina
had to visit a woman in her neighborhood who handed her the payment in
pesos, deducting a transaction fee. These material circulation networks not
only provided my research participants with cash but also facilitated access to
technical devices like the latest mobile phones, laptops, and LED ring lights
that are necessary for producing their videos but are not readily available in
official shops.

While the task of transmitting cash or foreign goods to friends and family
on the island is typically undertaken by members of the Cuban diaspora, these
informal practices have evolved into semiformalized systems that include private courier services and grassroots banking networks. In the United States,
informal bankers accept deposits that are then distributed by their employees
in Cuba (Tankha 2021). Additionally, the numerous travel and remittance agencies scattered throughout Miami and Hialeah also provide similar services, offering alternatives for those who cannot rely on personal networks of *confianza*
or kinship. However, personal networks, which AbdouMaliq Simone (2004)
describes as "people as infrastructure," often prove more useful in mitigating
economic hardships than formalized and business-operated alternatives. For
instance, following the government's elimination of the dual currency system,
which led to soaring inflation, Cuban Airbnb hosts and experience providers
encountered a significant dilemma. While the remittance service VaCuba was
converting U.S. dollars at a rate of only fifty Cuban pesos per dollar, the street

exchange rate had surged beyond eighty pesos. To avoid losses resulting from the poor exchange rate, Diego, a YouTuber who rented out his family's small apartment in the popular Vedado neighborhood, requested his friend in the United States, who managed his account, to accept Airbnb's payments on his behalf. The friend would then bring the funds in cash during his travels to Cuba, ensuring that Diego's family did not suffer financial setbacks due to unfavorable exchange rates.

Because YouTube pays higher advertising revenue for viewers from economically more developed countries (while viewers from Cuba generate no revenue due to the embargo), there is a strong economic incentive for Cuban content creators to target international audiences. Miguel, as an example, adopts strategies to gain more views and enhance his channel's popularity. One approach he takes is creating videos that deviate from his regular content. For instance, he produces cheesy supercuts featuring the most emotional moments from popular Korean soap operas like *Love Alarm*. By doing so, Miguel aims to capture the attention of fans of these shows who may not have a specific interest in content from Cuba but could stumble upon his channel. This diversification in his video content allows him to potentially attract a wider audience, generate more views, and increase his chances of monetization.

Successful Cuban YouTubers like Frank Camallerys and Anita con Swing have experienced organic growth in their audience by creating travel and tourism videos that showcase their country. Frank, in particular, noticed the lack of local content about Cuba online, which motivated him to create videos specifically tailored for Cubans in the diaspora. However, he discovered that his content resonated with a broader Spanish-speaking audience, including viewers from Spain and other Latin American countries. This realization prompted Frank to expand his content and produce travel videos featuring popular tourist destinations in Cuba. As Frank's YouTube channel gained popularity, he took advantage of the opportunity by launching a website that offered guided excursions to the destinations he featured in his videos, such as Viñales and Varadero. He also began receiving commissions from international travel agencies and online stores to promote their products. Being one of the first Cubans to turn social media into a profession, Frank initially faced skepticism and ridicule in a country where advertising was not yet widely prevalent. However, with the availability of mobile data in Cuba, local businesses started recognizing the value of Cuban influencers and began paying them to endorse their services or products. Dina, for example, found additional income by regularly promoting restaurants and beauty salons on her channel. Even for those who didn't amass as large a following as Dina, Anita, or Frank, the Cuban private sector's foray into online marketing opened up new opportunities for side jobs. Some of my interlocutors started working as social media or community managers for local businesses, often earning substantially better pay than their parents did working government jobs.

Popular Critique in an Emerging Public Sphere

As detailed in previous chapters, the Cuban government has the ability to exert control over the content distributed through vernacular infrastructures like el paquete and SNET. This control can be exercised by pressuring the individuals who operate these platforms or by imposing new laws and regulations. However, the government lacks the same level of control over the content that is circulated and expressed on global social media platforms, which are owned by U.S. companies and increasingly have (as I discussed in the previous chapter) become essential to Cuban daily life, a process that Plantin et al. (2018) have termed the "infrastructuralization" of digital platforms. This has led many young Cuban YouTubers to feel less pressure to self-censor compared to users and administrators of SNET or the compilers of the paquete semanal. They gradually realized the power of social media platforms and began pushing the boundaries of public expression. Facebook, Twitter, and YouTube thus have fundamentally transformed how Cubans can voice and engage in critique, significantly expanding public discourse. One example is Pedrito El Paketero, a highly successful Cuban YouTuber whose videos generate hundreds of thousands of views. He has become known for representing the voices of average Cubans in his channel. In his series "100 Cubanos responden" (100 Cubans answer), he interviews people on the streets about typically taboo topics. While some of his videos may employ provocative clickbait, such as "What's the strangest porno you have seen?" (Pedrito el Paketero 2019a), others, like "Do Cubans agree with their system? Are they happy?" (Pedrito el Paketero 2019b), genuinely invite his interlocutors (as well as his audience) to reflect upon their lives as citizens.

When I asked him if he considered his work as journalism, Pedrito explained that his goal was to disseminate the unfiltered opinions of individuals who are not represented in the official media. Some of the respondents in his videos are quite outspoken, which has occasionally resulted in backlash. Pedrito recalled an incident when he filmed two women on the street discussing their dreams of buying a house but being unable to afford it due to their low state salaries. As they described their living conditions, a plainclothes agent from state security approached them, identified himself, and began questioning them. The agent accused Pedrito and the women of engaging in counterrevolutionary activities and took them to the police station. After several hours of interrogation about the video's meaning, they were released without charges. Despite this incident, Pedrito remained unwavering in his choice of topics and even posted a series of cathartic video rants during the COVID-19 lockdown where he criticized the extreme shortages in the country and the high prices at government-controlled MLC stores. Pedrito was not the only young You-Tuber who faced backlash for making critical comments in their videos. Dina, for instance, lost her job at a state-run hotel after making a video accusing the Cuban police of inaction following a sexual assault she experienced while walking from the bus station to her mother's house at night. Miguel, her best friend, faced the threat of expulsion from medical school for creating a video that showcased the empty shelves in state stores.

However, the political stances of young influencers also extend beyond government criticism, and not all of their concerns conflict with the revolutionary ideology, which promotes principles like gender equality and resistance against U.S. imperialism and global capitalism (Kapcia 2008; Chase 2015). While the outspoken nature of young social media influencers has often led to conflicts with the state, there are instances where the government is uncertain about how to respond to them. In some cases, it has reluctantly started to incorporate the positions of a younger generation of activists, advocating for causes like animal rights or LGBTIQ+ rights (including YouTubers Jhans Oscar and Miguel Qbano from Dina's circle of friends) into the political process. For example, in April 2019, a march through Havana with around five hundred participants organized by young *animalistas* became the country's first authorized independent demonstration. In February 2021, the government then approved a long-awaited decree on animal welfare after a coordinated social media campaign by young activists exerted pressure on authorities. This was seen as a significant step for civil society as their demands were translated into law for the first time. However, the government remains wary of online youth activism, particularly when it spills onto the streets. This was evident when a pride march, spontaneously organized by gay rights activists through Facebook and WhatsApp after the government-controlled pride parade was unexpectedly canceled, led to the arrest of at least three participants. The government likely viewed the march as subversive, considering that a few months earlier, under pressure from evangelical churches, the National Assembly removed an article from the new constitution that would have legalized same-sex marriage, causing outrage among many young Cubans who supported it.[2]

These examples vividly illustrate the profound impact of the liberalization of Internet access and the mass uptake of social media on Cuban politics. It has brought about the establishment of a networked public sphere and the emergence of new platforms for discourse and civil activism. By examining the means through which political dissent was expressed before this liberalization, in the absence of a liberal public sphere as described by Jürgen Habermas (1991) and Nancy Fraser (1990), Marie Laure Geoffray (2015) identified three distinct "arenas of contention" characterized by different political positions and strategies. The first arena, which she refers to as the dissident arena, consisted of individuals who outright rejected the revolutionary utopia and criticized the political system as oppressive. However, these dissidents remained largely invisible within Cuba. The second arena, which she calls the critical arena, was linked to Havana's intellectual and cultural circles. It encompassed semipublic debates held in academic research centers such as the Instituto Cubano de Investigación Cultural Juan Marinello and select critical journals like *Temas*. The third arena, the contentious arena, comprised community activists and grassroots initiatives that operated at the margins and local levels. Geoffray describes how the lack of communication channels initially hindered connection and exchange between these isolated microspaces of debate. This dynamic began to shift only with the hesitant liberalization of Internet access and the rise of private ownership of computers and cell phones from 2008 onward. It resulted in a gradual transformation that paved the way for the creation of a transnational space of contestation, enabling the formerly disconnected arenas of debate to become interconnected and also incorporating voices and publics from the Cuban exile.

In the context of this newly interrelated arena of contention, two significant examples of collective action stand out: the Guerra de los Emails (email war) and the emergence of a Cuban blogosphere. The email war unfolded in 2007 as a digital debate on censorship among Cuban intellectuals both on and off the island. It was triggered by the airing of a favorable portrayal of poet Luis Pavón Tamayo on Cuban television, despite him being primarily remembered by Cuban artists and academics as the former head of Cuba's National Cultural Council during a period of heightened cultural repression in the 1970s. Initially, the debate started as an email exchange between a small group of peers, with their messages being forwarded to others. However, within a matter of days, the discussion expanded to include hundreds of contributions from various cultural practitioners circulating within Cuba and the diaspora. This marked the first instance of a semipublic and transnational discussion facilitated by digital media (for a comprehensive analysis of its impact and legacy, see Humphreys 2019).

Around the same period, activists began expressing their critical dissent through blogs, sharing alternative perspectives, criticizing government policies, and shedding light on social, political, and human rights issues within the country. Many of them faced severe consequences such as harassment, threats, and defamation (on the Cuban dissident blogger movement, see Henken 2011, 2017, 2021b; Duong 2013; Henken and van de Voort 2013; Vicari 2015). However, it is important to note that this new space of contention, which brought together previously separated arenas of critique through digital communication technologies, largely remained invisible to most Cubans outside of intellectual and activist circles. The vast majority of Cubans never learned about the email war or bloggers like Yoani Sánchez (who quickly rose to fame outside the island, winning human rights awards and engaging in dialogue with international celebrities) as the state maintained a monopoly on media, and the Internet remained inaccessible to most people.

Yet since the establishment of public Wi-Fi parks in 2015 and, more significantly, the introduction of mobile Internet in late 2018, the government has effectively lost its capacity to control and stifle contentious debates. As a result, it has, to a large extent, come to accept the proliferation of online public expressions of opinion. Cubans now actively engage in critical debates on platforms like Facebook groups, Twitter, and even the state news platform *Cubadebate*, something that would have been unimaginable just a few years ago. Today, it is not only young social media personalities like Pedrito, Dina, or Miguel who share critical videos about the ongoing economic hardships faced by the population and the government's inability to meet basic needs. Increasingly, ordinary individuals who do not aspire to be influencers or consider themselves dissidents also contribute to these discussions.

Facebook livestreams, in particular, have become a channel for everyday people to showcase their daily difficulties and make their voices heard when they do not receive institutional responses. These livestreams, which remain available as videos after the transmission ends, are often shared tens or even hundreds of thousands of times. One such instance occurred in June 2022 when Amelia Calzadilla, a young mother of three, recorded a *directa* (live video) that quickly went viral, gathering over 400,000 views and more than 7,000 comments. In the video, she passionately denounced the high cost of electricity in her home and the frequent power outages that adversely affected her family. She expressed frustration that she had not had a gas connection for ten years and that cooking with electricity was becoming increasingly unaffordable. Holding up documents to the camera, she demonstrated her unsuccessful appeals to various government bodies. In her moment of anger, she directly addressed Nicolás Liván Arronte, then the minister of energy and mines, and President Miguel Díaz-Canel, demanding that they prioritize serving the population.

The government's response to such public expressions of discontent is marked by a lack of confidence and uncertainty. In the case of Amelia, the government initially resorted to its usual tactics. She was subjected to ridicule on state television, accused in official press releases of being funded by foreign agents, and targeted with insults by the government-sponsored troll army known as *ciberclarias* (often composed of students from computer science institutes whose access to the Internet or cell phone connections depend on posting progovernment messages). However, as public support for Amelia did not wane, she eventually gained an audience at the government headquarters of her municipality, where she was allowed to express her concerns for over two hours.

The relative freedom that Cubans enjoy in this emerging virtual space of contestation can be attributed to several factors. Initially, the government largely ignored or failed to recognize the significance of this new space. More-over, the majority of Cubans who voice their opinions on social media plat-forms are cautious not to cross the delicate line between expressing criticism and directly challenging the system. For example, although Pedrito candidly addresses the daily struggles of life in his videos, he never denounces the Cuban socialist social and political order. Similarly, Amelia's intention in going live on Facebook was not to question the system itself but rather to hold government officials accountable as public servants and to pressure them to fulfill their promises of taking care of citizens. She made it clear that resorting to Facebook was her last option after her appeals to the authorities had fallen on deaf ears.

Another digital sphere where people came to express their grievances that quickly emerged with the mass proliferation of social media is the Cuban meme space. In Cuba, humor has always been a crucial survival strategy for navigating the many constraints of the nation's crumbling economy. Before the Internet, mocking politicians—or simply one's own problems—was confined to specific social microcontexts, such as networks of friends, where caricatures and graphic humor circulated. However, with the advent of mobile Internet, the dissemination of such forms of satire has exploded. Memes have rapidly inundated Cuban social media, ridiculing individual politicians, parodying an overwrought and repetitive party-state media discourse characterized by what Dominic Boyer and Alexei Yurchak (2010, 181) have described as "hyper-normalization" and critically highlighting facts that are omitted from official media channels. Virtually all my younger research participants consumed and shared clusters of memes, such as the "prayers to the god ETECSA" (which criticize the pricing and service quality of Cuba's national telecommunications company), the "Battle of Cuatro Caminos" (a series of memes related to the opening of a market in Havana where, due of the extreme shortage of prod-ucts offered there, multiple fights broke out between customers), or ostriches photoshopped into all sorts of mundane images (referring to Revolutionary commander Guillermo García Frías's suggestion during the TV news show *Mesa Redonda* that the problem of food shortages in Cuba could be solved by breeding ostriches instead of cows). The widespread presence of certain memes even has compelled state media to respond, thereby disrupting the previously one-way flow of political communication within the Cuban political system (González 2020).

The YouTube videos, Facebook livestreams, and memes discussed so far might not overtly challenge the Cuban political structure, propose alternatives, or incite resistance, yet they participate in an infrapolitical "hidden transcript" that nonetheless carves out new spaces for airing popular grievances. Therefore, while the increased availability of social media hasn't necessarily resulted in a heightened politicization of the population, it has expanded the scope for criticism and transformed the relationship between citizens and the state, fostering new channels for dialogue with state officials. The legitimacy of political authority continues to be largely accepted by the majority, yet the efficacy of its application for the public's benefit is increasingly coming under scrutiny.

Prior to the advent of the Internet, the government's handling of dissent was characterized by a division between the private and public spheres. As Hoffmann (2011) describes, criticism was allowed within private settings but was strictly prohibited from reaching the outside world, as encapsulated by slogans such as *bajo techo todo, en la calle nada* (under the roof everything, in the street nothing) or *la calle es de Fidel* (the street belongs to Fidel). While the government still seeks to retain control over the streets, as evidenced by its harsh response to the July 11 protests, it generally has come to tolerate critical expression within the new virtual public spaces primarily offered by U.S.-based digital platforms. This acceptance, however, is conditional, limited by two distinct boundaries. Individuals can express their critical opinions without retribution only as long as they resort to what digital politics anthropologist John Postill (2008, 419) calls "banal activism," that is, forms of engagement that are not quite political enough to pose a fundamental challenge to the system, and as long as they do so as private citizens and do not advocate for offline collective action.

The Emergence of Independent Journalism

Another discursive arena that emerged after the liberalization of the Internet was the multitude of independent digital news, information, and entertainment ventures that have surfaced since around 2015. These platforms provided new income opportunities for some of the young research participants involved in my study. Initially, independent journalism took the form of PDF magazines in the paquete distribution system (as described in chapter 2). However, with the increasing accessibility of the Internet, news websites like *Periodismo del Barrio*, *El Estornudo*, and *El Toque* have further contributed to the diversification of the Cuban public sphere. These outlets, often funded with the assistance of international institutions or the Cuban diaspora, have quickly replaced the Cuban blogosphere as the primary source of independent information about the country. Unlike blogs, which typically focus on limited topics and reflect the individual experiences and opinions of their authors, these news websites aim to provide competent, credible, and responsible journalism (Henken 2017; Geoffray 2021). They emphasize professional standards through attractive web design, frequent experimentation with multimedia storytelling, and engagement with readers via various social media channels. The contributors to these platforms are often young graduates from Cuban journalism programs who have become disillusioned with the state media system. While their education, such as at the Faculty of Journalism at the University of Havana, is surprisingly liberal, teaching the Western model of the press as the fourth estate and the ideal of balanced reporting, many of them quickly encounter frustration due to the limitations imposed on their work by the propagandistic state press (García Santamaria 2021; Natvig 2021).[3]

While the Cuban government considers independent journalism to be subversive and included it in the list of professions that must remain under government control in the 2021 reform of the private-sector regulations, most young journalists distance themselves not only from the official state-controlled media but also from the traditional opposition press, which primarily operates outside of Cuba and has limited reach within the island. They insist on their neutrality and objectivity, positioning themselves as observers rather than active participants in political resistance. Thus, their pursuit of independence from the government is less driven by political motives and more rooted in professional ethics and a desire for personal autonomy. Part of this personal autonomy includes the freedom to explore economic opportunities in the private sector. Many of the young independent journalists I encountered possess a significant entrepreneurial spirit. For example, my friend Bryan, who is proficient in English, leverages his language skills to generate additional income by occasionally publishing analysis and opinion pieces in international news outlets. He achieves this, for example, by cold-emailing editors on the business networking platform LinkedIn.

One of the notable success stories in the realm of young Cuban news entrepreneurs is José Jasán Nieves Cárdenas, the founder of *El Toque*. His career began during the political thaw of the Obama period when the Cuban government allowed some international media platforms, including *Contemporánea*, *Progreso Semanal*, and *OnCubaNews*, to have correspondents in Havana. These outlets, managed by Cubans in the diaspora and catering to a diaspora audience, started hiring local journalists, thereby creating an alternative job market where contributors were paid in the dollar-pegged CUC, resulting in significantly higher salaries compared to those offered by the official media. It was during this time that Nieves Cárdenas became involved with *El Toque*, which was initially a digital web platform run by the Dutch NGO RNW Media and focused on human rights and freedom of expression issues in Mexico, Cuba, and Venezuela. In 2016, when RNW Media decided to withdraw from Cuba, Nieves Cárdenas, then twenty-nine years old, successfully negotiated to keep *El Toque* alive as an independent news platform operated and managed from within the island. To facilitate the transition from its Dutch founders, he had to navigate various challenges. He had to establish a new legal status for *El Toque* (ultimately registering it as a foundation in Poland) and secure new international funding sources. Additionally, he faced the critical task of finding a bank willing to facilitate the transfer of funds to Cuba despite the U.S. embargo. Today, *El Toque* stands as a remarkable achievement and one of the most renowned independent news organizations on the island. It keeps the state press on its toes through its selection of topics and innovative use of multimedia formats.

While ordinary Cubans can express critical views online within certain limits without facing significant backlash, the Cuban state systematically harasses and periodically blocks independent online news outlets. This oppressive environment became particularly unbearable for Nieves Cárdenas when he was placed under house arrest by state security on the day of the planned SNET demonstration in June 2019 (as described in chapter 3). Shortly after this incident, Nieves Cárdenas and his partner Elaine Díaz, who is the founder of *Periodismo del Barrio*, traveled to a conference in the United States, where they received news that Elaine was pregnant again. It was at this point that they made the decision not to return to Cuba, as they were unwilling to raise their children amid ongoing repression. Since that day, *El Toque*, like many other Cuban journalism platforms, has transformed into a transnational media operation. It is currently produced from seven countries, with its core team scattered across Cuba, the United States, Canada, Mexico, Spain, Italy, and Mexico. It relies on outside assistance for funding, web hosting, and legal registration, while also frequently witnessing the departure of employees and freelancers from Cuba in search of new opportunities overseas, such as scholarships to pursue studies in Europe, the United States, or Mexico. Operating from Miami has brought some advantages for *El Toque*. It has simplified financial transactions since funds no longer need to be funneled into Cuba, bypassing the U.S. embargo. Additionally, the emergence of new donors, including Cuban American philanthropists, has provided a welcome source of support. However, this shift has also introduced new dilemmas for Nieves Cárdenas and his team.

Between the Front Lines

Independent journalistic ventures like *El Toque* and *Periodismo del Barrio* face the challenge of maintaining their independence not only from the Cuban state but also from the influence of the U.S. government. Like their Cuban counterparts, U.S. government bodies view the existence of nonofficial media in the country as a form of dissidence. Institutions such as the U.S. National Endowment for Democracy (NED) and the U.S. Agency for International Development (USAID) have provided significant financial support over the years to build a Cuban oppositional movement. These entities are therefore particularly interested in media-savvy social media influencers and independent journalists from the island. *Periodismo del Barrio* openly addresses the dilemma by disclosing the funding they receive. Their code of ethics states that they accept collaboration, whether in terms of financing, equipment, or services, only from organizations, institutions, foundations, and individuals who respect the sovereignty of the Cuban state, uphold the platform's editorial independence, and are transparent about their identity, their purposes, and the origin of their funds. This approach sets them apart from other news websites claiming to be Cuban but operating outside the island, such as ADN Cuba and Cubanet, which are backed by U.S. institutions with the explicit goal of regime change in Cuba and openly advocate for democratic transition. Much of the reporting on these sites tends to be sensationalist and primarily relies on social media content from the island, often given a particular political spin. They do not maintain a network of professionally trained correspondents on the ground.

Funding from various U.S. institutions enters Cuba through different channels, including grants, journalism awards, and NGO projects, and not all donors have the explicit goal of overthrowing the Cuban government or building an internal opposition. The U.S. government has pursued a two-pronged strategy in Cuba, as well as in other countries like Venezuela. On one hand, they provide financial support to the political opposition, and on the other hand, they assist young professionals and independent businesses in order to bolster an emerging middle class that they hope will play a significant role in future transition scenarios.

For journalists, intellectuals, and artists in Cuba, especially those who need to maintain some connection to the system for economic reasons, navigating acceptable funding sources becomes a delicate balancing act. This is particularly challenging considering that the official state media readily labels all recipients of U.S. funds as mercenaries and pawns of foreign powers. A writer friend of mine, for instance, regularly contributes to several online journals on Cuban art, literature, and cinema that receive grants from the United States, while also holding a government job at a literary education center. As long as the websites he writes for are accessible in Cuba and he avoids touching on political issues, his boss accepts his sideline. However, it would be impossible for his superior to write for these websites herself.

For artists and writers who are considered dissidents by the Cuban government and are therefore prohibited from publishing or working within the state-controlled cultural sector, the money flowing into the country through these platforms often becomes their sole source of income. This situation makes it difficult for them to maintain a critical distance from both systems. As an actress and writer friend sarcastically remarked to me, with "half of Havana's intellectual scene kept afloat by U.S. anti-Castro money," she feels "beset by two mafias."[4]

According to investigative journalist Tracey Eaton (2021) and his Cuba Money Project, many U.S. democracy promotion efforts specifically target aspiring young Cubans through scholarships and international workshops offered by opaque NGOs, often based in other Latin American countries. Such initiatives often emphasize concepts such as "young leadership," "human rights," and "civil society," making it challenging for applicants to discern the underlying interests. Many of my young research participants, for example, eagerly took part in a contest presented by Cuban American social media personality Alexander Otaola on his web show *Hola Ota-Ola* that aimed to discover and support the next YouTube stars from Cuba. During the show, Otaola claimed that the contest was financed by "collaborators in Mexico." Participants were required to create short videos on topics such as "dreams of a future Cuban society" or "life in my neighborhood." Winners were promised new computers, smartphones, international travel opportunities, and 10 GB of monthly mobile data for an entire year. Those selected also received an intensive online program with courses on digital privacy and audience building. They were further supported in developing a social awareness campaign of their choice, with additional funding available for printing stickers and buttons and producing polished social media content. Despite none of the program participants I knew were openly challenging the socialist order in their campaigns, I had a sense that the anonymous sponsors considered the outcomes to be a success.

Otaola, an entertainer and self-proclaimed anticommunist activist, gained fame through his YouTube channel where he satirically comments on politics, news, and celebrity culture in Cuba and the Miami exile community. His lengthy webcasts mainly consist of tirades against the Cuban state as well as criticism toward Cubans both on and off the island who he believes do not sufficiently speak out against the government. Otaola has been highly effective in organizing boycott campaigns against Cuban artists and musicians touring the United States, accusing them of colluding with the Cuban system, for example, Haila Mompié and Gente de Zona. Alongside other influencers such as Eliécer Ávila, Michelito Dando Chucho, and Ultrack, Otaola is considered a driving force behind Cuban American support for Donald Trump in recent years. The influence of social media sources, including these influencers, on Cuban Americans' political viewpoints is significant. According to a poll conducted by Florida International University in 2022, 37 percent of respondents reported receiving information about Cuba from social media and approximately 19 percent acknowledged that social media influencers do influence their political attitudes to some degree. This is even more pronounced among the younger respondents and recent arrivals (Florida International University 2022, 45–46). Consequently, these nonconciliatory views on U.S.-Cuba policy have contributed to the growing political polarization within the Cuban American community, leaving little room for nuanced perspectives.

After relocating to the United States, Nieves Cárdenas and his partner Elaine Díaz have become the latest victims of Otaola's defamation campaigns. Ironically, while the Cuban state media have often grouped the three together as "counterrevolutionary puppets of U.S. interests," Otaola has singled out the couple on his show for their rejection of the U.S. embargo and their advocacy for dialogue between both governments. He has repeatedly referred to them as "pallid critics" and "ex-oficialista," highlighting that they both (like virtually all Cubans on the island) at some point worked for Cuban state institutions. These accusations were subsequently amplified by platforms like Cubanet and other U.S.-based websites, resulting in Nieves Cárdenas and Díaz receiving personal threats through social media. Reflecting on the experience of being caught between two ideological fronts, Nieves Cárdenas shared with me that at times he felt more fear in Miami than in Cuba. In Cuba, the repressive state system was more predictable, and the path to physical harm, such as beatings, imprisonment, and even death, was typically long and gradual while in the United States a "madman with a gun" could lurk behind every corner. Several of my contacts, both inside and outside of Cuba, who have also been targeted by Otaola (including DinaStars, Pedrito El Paketero, and Michelito Dando

Chucho) likened his character assassination campaigns to the infamous *actos de repudio* (acts of repudiation) organized by the Cuban state that involve mobilizing loyalist mobs to protest outside the homes of dissidents, verbally assaulting and shaming them.

Spreading anticommunist messages on social media has become a profitable business model in Miami, where young exiles are quickly drawn into what Alejandro Portes (2007) referred to as the "Cuban-American political machine." With their media-savvy skills, newcomers can easily find employment with propaganda media outlets directly or indirectly financed by the U.S. government, such as *Radio y Televisión Martí* or Cuban American news websites. As they immerse themselves in this environment, they often feel compelled to conform to the dominant views of the community. Otaola's successful recipe, blending comedy, gossip, and right-wing ideology, has inspired other social media personalities like Los Pichy Boys, who were primarily known for their apolitical humor until a few years ago. This recent radicalization trend has also impacted the transnational public sphere, which emerged during the thaw between the United States and Cuba, as described by Alberto Laguna (2017) in his account of Cuban diasporic comedy. While both Los Pichy Boys and Otaola, before his radicalization, were embraced by Cubans on the island as well as in Florida, enjoying a substantial presence in the paquete semanal, the compilers had to remove their videos once their content became more politically oriented. Consequently, new ideological boundaries have formed between the emerging "dual audiences" on both sides of the Straits of Florida.

Online Protest Movements: From San Isidro to 27N

The growing capability of Cubans to forge connections through social media based on mutual interests and concerns has sparked the creation of novel civil society initiatives. As explored extensively in this book, such digital interconnectedness has played a crucial role in a variety of collective actions. The resistance to the SNET shutdown outlined in chapter 3, the self-organizing methods on messaging platforms such as WhatsApp and Telegram discussed in chapter 4, and the protests spearheaded by animal rights and LGBTIQ+ activists, which this chapter details, signify an intensifying dynamic of engagement between the state and emerging groups of autonomous activists. While these activists do not represent the large majority of Cubans who refrain from openly disputing the legitimacy of the political system, some of these groups have started to directly confront the government. Their tactics extend beyond online mobilization to include public demonstrations, signaling a new chapter in the expression of civil discontent.

One notable group that gained significant attention is the Movimiento San Isidro (MSI), which was established in late 2018 by artist-activists in response to the government's Decree 349. This decree aimed to increase control over the growing independent artistic community in Cuba. In November 2020, a member of MSI, rapper Denis Solis, was sentenced to eight months in jail for "contempt" after sharing his encounter with a police officer who had invaded his home on social media. In solidarity with Solis, several members of MSI initiated a hunger strike at artist Luis Manuel Alcántara's home. However, when state security agents forcibly evicted the hunger strikers, citing COVID regulations, a spontaneous protest erupted in front of the Ministry of Culture. This demonstration, which began with only a few participants on the morning of November 27, 2020, quickly spread through social networks and messaging applications. By evening, hundreds of young artists and intellectuals had gathered (including many friends of mine who had previously refrained from political activism). The protesters demanded a meeting with the minister of culture, Alpidio Alonso, to initiate a dialogue on censorship and freedom of expression. Caught off guard, the government initially succumbed to the pressure and agreed to engage in negotiations. With the intervention of renowned Cuban filmmaker Fernando Pérez and actor Jorge Perugorría, the deputy minister of culture, Fernando Rojas, agreed to meet with a delegation representing the protesters.

The Grupo del 27N (27N Group), which emerged as the representative body for the protesters, encompassed a wide range of political viewpoints. Many of its members had previously been involved in the state-funded cultural sector, and not all of them shared the MSI's antigovernment position. According to Cuban sociologist Rafael Hernández, who is acquainted with many of its members and conducted interviews with several of them for an article published in *Nueva Sociedad* (2021), the group faced challenges in reaching a consensus. Following days of deliberation, the radical faction within the group moved forward by sending an email outlining new demands, including a request to negotiate directly with President Díaz-Canel. In response, the Ministry of Culture terminated the dialogue, citing the introduction of unacceptable conditions that exceeded the agreements made on the night of November 27. The state then employed its familiar strategy of denouncing the protesters as mercenaries and foreign agents through official media channels. Additionally, it intensified repression against independent journalists, including measures such as cutting off their mobile Internet access and summoning them for questioning.[5]

On January 27, 2021, exactly two months after the initial protest, young activists gathered once again outside the Ministry of Culture on the symbolic day of José Martí's birthday. Their objective was to revive a dialogue with the government and advocate for the release of detained artists and journalists who had been apprehended earlier that morning. In an attempt to avoid the optics of another protest in front of the ministry, officials initially invited the small group into the building, but with the condition that they leave their cell phones outside. When the activists refused this request, tensions escalated, culminating in an incident where the minister of culture himself forcefully slapped the cell phone out of the hands of an independent reporter who was filming him. Despite the events of the day concluding with the mass arrest of artists, activists, and journalists, and the government implementing a nationwide Internet blackout, photos and videos capturing the episode rapidly spread across the Internet in the days that followed. The act of Alpidio Alonso slapping the journalist's cell phone became a powerful image, as Ted Henken (2021c, 78) aptly described, representing the government's disdain for independent sources of information (and, naturally, it was quickly turned into a meme). The widespread circulation of a video showing the minister's attack also underscored the irreversible loss of the state's monopoly on communication.

Another viral video that had an immense social and political impact was the music clip for the song "Patria y Vida" (Homeland and life), released in February 2021. The video reached virtually all of my research participants within a day of its release. It featured performances by highly successful exile artists such as Yotuel Romero (based in Spain), Gente de Zona, and Descemer Bueno (residing in Miami), along with Cuba-based rappers Maykel Osorbo (a member of MSI) and Eliécer "El Funky" Márquez, who both had played significant roles in the escalating events in Havana. The song's title directly challenged Fidel Castro's revolutionary slogan "Patria o Muerte" (homeland or death), and its lyrics are a deft critique of the Cuban government's ideological obstinacy and divisive rhetoric. The video resonated profoundly on the island, appealing to a population that had grown weary of constant sacrifices. Eventually, Patria y Vida became the slogan of the protests that erupted on July 11, shaking the entire country. It is a fascinating cultural text that has generated numerous contradictions through its reception and dissemination, which I will unpack in the following discussion. The song and video aimed to reclaim the mobilizing power of patriotism from the state's exclusive control, as the government had historically relied on stirring up nationalist sentiments against external adversaries to maintain its grip on power. The video prominently featured performance artist Luis Manuel Alcántara, an unofficial leader of MSI, striking a patriotic pose with the Cuban flag, as well as footage from the 27N protests and the waves of solidarity they generated among the Cuban exile community.

Much of the video's impact stems from the fact that it exclusively features Black artists who, despite their humble origins, have achieved Grammy-winning success. This representation resonated strongly, particularly among Cuba's Black population, which had borne the brunt of the economic hardships in recent years and had not reaped significant benefits from Raúl Castro's economic reforms. Consequently, it is not surprising that the July 11 protests, during which the song became an unofficial anthem, primarily erupted in Havana's marginalized Afro-Cuban neighborhoods. Ironically, the song also found support within Florida's predominantly white and Republican exile community, which typically rejects discussions about structural racism in the United States or within their own community (Bustamante 2021). During my research stay in Hialeah, which took place a few months after the July 11th protests, the city was still awash in Patria y Vida banners and the artists were celebrated akin to war heroes within the Cuban American community.

Just a year prior to the release of the song, Gente de Zona had become the subject of one of Otaola's boycott campaigns after publicly praising Miguel Díaz-Canel. As a result, they were excluded from a New Year's Eve concert in Miami following the vocal opposition of the city's Cuban American Republican mayor, Francis Suarez. A few months later, they gave in to the pressure and started to rebuild their career on the lucrative U.S. market. They performed at the Free Cuba Fest, under a large screen that displayed messages such as *abajo el comunismo* (down with communism), exemplifying once again the influence of the Cuban American political machine in aligning individuals with its agenda. The Cuban state media responded to Gente de Zona's change of stance with a barrage of attacks and insults, including homophobic, racist, and elitist remarks, reiterating their long-standing criticism of reggaeton music, which always had such undertones. However, the artists themselves arguably compromised the powerful antiracist message of their song by catering to Miami's ultraright exile media. Patria y Vida was treated as a public relations success by these outlets, as well as by government officials like USAID director Samantha Power, who praised it on Twitter.[6] USAID, known for its interest in targeting Cuba's hip-hop scene as potential agents of opposition, even made explicit reference to the song in a funding appeal of two million dollars aimed at supporting groups that utilize culture to foster social change in Cuba. The agency stated that the song "not only created greater global awareness of the plight of the Cuban people, but also served as a rallying cry for change on the island" (USAID 2021).

July 11: The Escalation

The technological, social, and political transformations that I have examined throughout this chapter and this book as a whole have culminated in an event that represents a significant turning point in recent Cuban history: the eruption of public discontent on July 11, 2021. A combination of multiple factors coincided to create a perfect storm: the increased availability of Internet access, which eroded the state's monopoly on information, the tightening of the U.S. embargo under the Trump administration, the devastating impact of the COVID-19 pandemic on Cuba's tourism industry, the growing dissatisfaction among young activists and artists regarding political stagnation, and the economic disruptions caused by the untimely currency unification. These broader factors intertwined with local triggers, including the collapse of an electricity plant in the city of Holguín and a deteriorating health situation in the city of Matanzas.

In the week leading up to July 11, the province of Matanzas, located two hours east of Havana, was on the brink of a public health crisis. The region experienced an uncontrollable surge in COVID-19 cases, mostly imported by tourists from the nearby beach destination of Varadero, which was exacerbated by severe shortages of medicine, food, and hygiene products. News about the dire situation quickly spread across the country, thanks to the efforts of activists who utilized hashtags such as #SOSCUBA and #SOSMATANZAS to raise funds, gather medical supplies, food, and other essential items, and deliver them to the affected areas. Over time, this grassroots initiative gained support from civil society organizations, governmental bodies, and Cuban exiles residing in the United States, Spain, and various Latin American countries. Some individuals even called for a "humanitarian corridor" to address the pressing health emergency, a proposal that the Cuban government vehemently rejected, viewing it as foreign interference. The online campaign gained further momentum through the backing of both internal and external opposition groups within Cuba.

In addition to the escalating COVID-19 crisis, the profound social and economic problems Cuba was grappling with had become increasingly unbearable for its citizens. Widespread shortages, soaring inflation, and frequent power outages contributed to a growing sense of frustration among the population, pushing many to their breaking point. The initial protests on July 11 originated in San Antonio de los Baños, a small town located just outside Havana. According to an investigation conducted by the independent news platform *El Estornudo* (Colomé 2021), which was even covered by official Cuban state media, the demonstrators responded to a call that was published in a local

Facebook group administered by two individuals using pseudonyms, along with an exiled Cuban residing in Miami. While this trio had made previous attempts to mobilize people from the town for protests, their message struck a chord this time. Videos of protesters chanting "Patria y Vida" and shouting "Libertad" (Freedom) were livestreamed on Facebook and quickly shared across social media platforms, garnering attention from independent Cuban media outlets as well as Miami-based opposition media.

The news of the protests quickly spread and ignited similar demonstrations in numerous other cities throughout Cuba, including Camagüey, Holguín, Matanzas, and Havana. In response to the escalating situation, President Díaz-Canel hurried to San Antonio de los Baños to engage in dialogue with the protesters. Subsequently, he appeared on state television, interrupting the broadcast of the European soccer championship final, which was being widely watched by Cubans, and called upon government supporters to take to the streets in opposition to the protesters. Additionally, the government resorted to cutting off access to the Internet in an attempt to disrupt the demonstrators' ability to organize.

I had departed Cuba three weeks prior to July 11, concluding a period of seven months of fieldwork. Throughout my time there, I witnessed the mounting frustration and toll caused by the government's stringent lockdown measures due to the pandemic. However, I was taken aback when my contacts started sending me videos of the protests unfolding in San Antonio de los Baños. Concerned about the safety of my friends who worked at the esteemed international film school in the town, I attempted to reach out to them, but the intermittent Internet outages hindered communication. In the ensuing days, I found myself glued to the screen, diligently following the unfolding events through social media and the Cuban independent press.

One of my most active Cuban contacts was Dina, who utilized Psiphon, a VPN that gained significant promotion within the transnational Cuban social media sphere during the protests, to continue reporting and sharing information about the situation on Instagram and Twitter. Two days later, on July 13, she uploaded a video to YouTube showcasing her and her friends participating in the protests. On the same day, she joined a panel discussion on the Spanish television program "Todo es Mentira" (Everything is a lie) via Zoom, alongside Yotuel, one of the performers of the song "Patria y Vida." While Dina was live on air, Cuban state security officers arrived at her home, taking her away for questioning right in front of the Spanish presenter and the national television audience. After not hearing from Dina for twenty-four hours, her friends initiated a social media campaign to raise awareness about her arrest and urge people to inquire about her whereabouts. This spontaneous organizing through digital platforms was a skill developed by the generation of connected Cubans at the heart of this chapter, honed through previous civic activism campaigns. Adriano, in particular, tirelessly encouraged other Cuban YouTubers to speak out in support of Dina, mobilizing crowds outside various police stations where she was suspected to be held captive. He also provided comfort to her mother, who even posted a video herself, expressing her fear and concern. The hashtag #FreeDina quickly gained traction among numerous Latin American social media influencers, drawing the attention of news media worldwide (this escalating series of events is documented in my documentary *Dinamita* that accompanies this chapter). Upon her release, Dina promptly went live to assure her supporters that she was well. Visibly affected by her encounter with state power, she subsequently distanced herself from further political activism in the following days.

The protests that erupted on July 11 were ultimately short-lived, as the state quickly regained control of the streets through the deployment of military police and plainclothes security guards armed with weapons such as wooden clubs and metal bars. Although smaller protests emerged in the following days, they were swiftly dispersed due to the significant police presence. Hundreds of demonstrators were arrested, and many of them faced lengthy prison sentences in the subsequent months. On July 17, the government organized a counter-rally, mobilizing tens of thousands of people, indicating that there were still many Cubans willing to respond to the state's calls for mobilization. Commentators both inside and outside of Cuba identified the younger generation and Afro-Cubans from marginalized neighborhoods as the driving forces behind the protests. Harold Cárdenas (2021), the founder of the online journalism platform *La Joven Cuba* (The young Cuba), highlighted a "generation gap" in Cuban politics, where young people are excluded from positions of power and therefore find their sources of identification not in the traditional political system (that for decades was based on charismatic leadership) but rather on social media. Similarly, my friend Bryan Campbell Romero (2021) examined in an opinion piece for the website of the North American Congress in Latin America (NACLA) how the experiences of Black Cubans are consistently overlooked by the Cuban government. However, interpretations of the protesters' goals varied depending on ideological perspectives. While the protests were largely seen as a spontaneous expression of frustration regarding food and medicine shortages, power outages, and curbs on civil liberties, there was no clear consensus among observers on the demonstrators' specific demands or objectives. While Cuban American opposition media interpreted the protests as calls for political change, the messages and demands voiced by the protesters did not coalesce into a singular, defined goal. Rather, they expressed anger over various grievances for which viable solutions were not readily apparent.

In response to the concerns raised during the manifestations, the Cuban government took some steps to address the issues. It implemented measures such as allowing unlimited and tax-free imports of food, hygiene items, and medicines for private use until the end of the year. It also fast-tracked the implementation of certain overdue economic reforms for private businesses. Additionally, the official press demonstrated a heightened concern for citizens living in economically deprived circumstances, reflecting a recognition of the legitimate complaints expressed during the demonstrations (Romay Guerra 2021).

However, in addition to these actions, the government responded by escalating political pressure on online spaces through the implementation of Decree 35. This law prohibits the dissemination of false news and incitement to violence, with the definition of false news determined by the state.

The role of outside interference in the July 11 protests in Cuba is a matter of contention among political analysts. While it is evident that social media and the recent expansion of Internet access in Cuba played significant roles in facilitating the spread of criticism and mobilization, opinions differ regarding the influence of external actors. President Díaz-Canel has repeatedly claimed that the protestors were confused and manipulated by coordinated online campaigns orchestrated from abroad. *El Estornudo*'s investigation revealed that one of the individuals behind the San Antonio Facebook group, which sparked the initial protests, is based in Miami. Spanish social media analyst Julián Macías Tovar also highlighted in a series of tweets that the hashtag #SOSCuba originated in Spain and received substantial amplification through automated retweets from newly created accounts.[7] These automated accounts were also responsible for spreading fake images and videos to exaggerate the size of the protests. Some of these included photos taken during the Arab Spring protests in Egypt that were falsely presented as demonstrations on the Malecón in Havana. Additionally, progovernment gatherings were mislabeled as antigovernment protests. However, even if we acknowledge the existence of a large-scale, coordinated social media campaign originating from outside Cuba, it is difficult to assess its actual impact. Carlos Alzugaray (2021) accurately highlights that these protests, even if triggered externally, would not have escalated to the extent they did if they hadn't found a receptive environment due to various political mistakes and mismanagement by the government.

After the Protests

In the months following July 11, a new citizen initiative called Archipiélago (Archipelago) emerged, with the aim of organizing a protest march on November 15 (15N) through social media. The march was intended to demand the release of the July 11 prisoners, respect for human rights, and a democratic and peaceful dialogue with the government. Archipiélago was founded on August 9, 2021, as a Facebook group and quickly gained over thirty thousand online members. Playwright Yunior García, known as one of the more moderate representatives of the 27N movement, led the group. Archipiélago spoke out against the U.S. embargo and foreign interference and even requested permission from authorities for its proposed march based on an article in the new Cuban constitution, which guarantees the right to peaceful demonstration. Nevertheless, the government tried to prevent the march by any means necessary. State security harassed and interrogated individuals expressing their plans to participate, while launching an intense state media campaign to discredit Yunior García as a U.S. mercenary. When the government banned the march, Yunior had announced that he would walk alone through Havana, dressed in white and holding a white rose, alluding to a José Martí poem in which the protagonist offers such a flower to both a sincere friend and a cruel tormentor.[8] On the day of the planned protest, government forces and sympathizers surrounded the homes of Yunior and other prominent Archipiélago leaders to prevent them from taking to the streets. Although government supporters partially obstructed his window with a large Cuban flag, a journalist managed to capture a photo of Yunior extending a white rose to the viewer. This iconic image came to represent the Marcha Cívica por el Cambio (Civic march for change) that ultimately never took place.

Two days later, Yunior and his partner Dayana Prieto, who was also active in the group, made a sudden appearance at Madrid airport after fleeing the country on a tourist visa with the assistance of the Spanish embassy. Their decision to leave Cuba out of fear of prosecution came as a shock to many of my friends and interlocutors who sympathized with the movement. While most expressed understanding for Yunior's personal choice to prioritize his freedom, there was a sense of betrayal among some who felt abandoned by his quiet departure.

The absence of their prominent leader left a void within the Archipiélago group, and as a result several other organizers resigned, leading to the rapid dissolution of the movement. This development left many of my friends and acquaintances, who belonged to the digital millennial generation, feeling disheartened. Initially hopeful for political change following the protests and the emergence of Archipiélago, their optimism waned as they witnessed the government's readiness to suppress the growing opposition through violent means. Moreover, the departure of the seemingly sole potential opposition leader further contributed to a sense of apathy and disillusionment among them.

The political paralysis following the events of November 15, coupled with the worsening living conditions in Cuba, led to an unprecedented wave of emigration, particularly among the skilled and ambitious younger generations. In 2022, the number of Cubans leaving the island surpassed even the combined figures of the Mariel and Balsero crises, the two significant migration crises Cuba had experienced in 1980 and 1994.[9] According to data from U.S. Customs and Border Protection, almost 353,000 Cubans entered the United States between September 2021 and February 2023. This represents a staggering 3.5 percent of Cuba's total population and between 5 and 6 percent of its active labor force (Grenier 2023).

The Cuban government, in an effort to alleviate political pressure and address the growing discontent within the country, played a significant role in facilitating the out-migration of its citizens. In November 2021, the Nicaraguan government, a close ally of Cuba, took a significant step to assist the Cuban regime by allowing Cuban citizens to enter Nicaragua without requiring a visa. This decision created a new migration route from Cuba to the United States through Central America. This "export" of migrants (and the subsequent "import" of their family remittances) is a long-standing government strategy to stabilize the political order (Hoffmann 2016).

The United States continues to grant special privileges to immigrants from Cuba, distinguishing them from immigrants of other nationalities. This preferential treatment has its roots in the historical use of immigration policy as a tool for foreign policy objectives and as a means to showcase the superiority of the capitalist system to Cubans (Eckstein 2022). While the entitlements enjoyed by earlier waves of Cuban migrants have been scaled back by successive administrations, such as under Obama's efforts to normalize relations with Cuba and Trump's overall anti-immigration policies, the fundamental framework of the Cuban Adjustment Act remains intact. Under this act, Cubans who are inspected and admitted or paroled into the United States are eligible to apply for a permanent resident permit after one year. This allows them to bypass standard immigration rules and obtain legal status in the United States. Furthermore, upon arrival in the United States, Cubans who are paroled may apply for employment authorization and welfare benefits.[10]

Cuban migrants in search of better opportunities have turned to not only the United States but also Spain. The Ley de Memoria Histórica (Law of historical memory), implemented in 2007 and subsequently enhanced by the more liberal Ley de Memoria Democrática (Law of democratic memory) in 2022, has provided a pathway for thousands of Cubans to obtain Spanish citizenship if their grandparents were exiled from Spain during the civil war and the dictatorship. Furthermore, many talented young Cubans have successfully pursued opportunities in Spain through scholarships, artist residencies, enrollment in Spanish universities, and participation in conferences, lectures, and exhibitions, often overstaying their visas. The Spanish authorities have also shown generosity in granting asylum status to dissidents and individuals involved in recent waves of protests, further encouraging Cuban migration to the country.

Those who are able to leave Cuba through these means are typically people who have the financial resources to fund their migration or have support from family members abroad. Many of the newly arrived migrants I spoke with in Miami mentioned spending substantial amounts, up to ten thousand dollars, on flights to Nicaragua and paying human smugglers who assisted them in crossing the U.S. border through Mexico. Similarly, Spanish authorities require proof of economic self-sufficiency, such as eight thousand euros in a bank account, for issuing student visas. These requirements limit the opportunities for disenfranchised Cubans to leave, who were among the primary participants in the July 11 protests and are the most affected by the country's economic crisis, as they often lack transnational personal networks that can provide financial support or remittances. Consequently, the current wave of migration represents an unparalleled brain drain, predominantly composed of young, educated, and ambitious Cubans. This is likely to mortgage the country's future, as Cuban migration researchers at the Centro de Estudios Demográficos (CEDEM, Center for Demographic Studies) have warned (Aja Díaz and Rodríguez Soriano 2022).

Of my friends and acquaintances whose experiences I have described in this chapter, the majority have left Cuba, and many have resorted to selling their possessions in order to escape. Frank, Anita, and Miguel have all migrated to Spain, while Bryan is now in Mexico. David has gone to the extent of selling his parents' house, hoping to secure a *visa no lucrativa* (nonprofit visa) for Spain to bring his partner and entire family along. Among Cuban YouTubers, there has emerged a new genre of videos documenting the journey from Nicaragua and the crossing of the U.S. border. Emma Style, Adrian Peachy, and even Frank Camallerys's mother (on her son's channel) have captured their experiences on this treacherous route. Emma, who won Otaola's YouTuber contest and subsequently faced threats from Cuban state security, has even shared a highly detailed video offering tips and guidance for preparing for the passage (EmmaStyle 2022).

All of the main protagonists of my documentary have left Cuba as well. Jhans had already departed for Moscow in 2019 with his Russian partner. Adriano, who holds a Spanish passport through the Ley de Memoria Histórica, overstayed his tourist visa during a trip to the United States and now resides in Hialeah. Dina, on the other hand, successfully secured a one-year program in entrepreneurship and business innovation at a private university in Madrid, where she is part of a cohort predominantly composed of Cuban students. Despite the option of seeking political asylum in Spain following the events of July 11, Dina chose to save enough money to come as a student because she wanted the opportunity to return to Cuba and see her mother. When I visited Adriano in Hialeah, he was close to completing the one-year requirement before he could apply for permanent residency. Although he couldn't open a bank account without proper documentation, he had started building a livelihood by relying on the diasporic support networks in the city. He worked as a video editor for a Cuban-run beauty clinic, rented a small efficiency apartment from a Cuban landlord, and received his salary and paid rent in cash. He even found a way to use Amazon and Uber without a credit card by purchasing gift cards from drugstores or convenience stores. Similar to Dina, he made the decision to steer clear of the highly polarized transnational Cuban media sphere and opted not to pursue a social media career as a *contrarevolucionario*. He expressed that the first thing he wanted to do after receiving his residency papers was to travel to Cuba.

Conclusion

Mobile Internet access proved to be a transformative tool for young, tech-savvy Cubans, empowering them to utilize social media for economic advancement, critical discourse, and civic activism. The platformization of Cuba's authoritarian public sphere has so far resulted in fewer of the political challenges that communication scholars have identified in liberal societies when corporate entities begin to dominate the public sphere, such as echo chambers, capitalist-driven surveillance, and the amplification of sensationalist, polarizing, or misleading content. While Facebook was overwhelmed with misinformation from various ideological factions during the events of July 11, the state still stands as the primary disseminator of disinformation, exercising control over mass media to primarily broadcast messages that serve to legitimize itself. Besides state-harassed independent journalists, the state-controlled Cuban public sphere has lacked the intermediaries who take concerns from society, process them, and formulate policy issues that Habermas and others consider necessary for the proper functioning of the deliberative public sphere—a role that, as I have demonstrated in this chapter, some Cuban netizens on Facebook and YouTube have now assumed for themselves.

For profit-driven corporations such as Facebook, Google, and Twitter, Cuba is not a relevant market as these companies are effectively unable to profit from data extraction or user engagement or sell advertisements on the island, due to the U.S. embargo and Cuba's isolation from international financial and goods markets. My young YouTuber friends have experienced this firsthand as they, despite their popularity, are unable to monetize their viewership in Cuba through the platform. This economic irrelevance has prompted some of these platforms to overcomply with the complex and occasionally contra-dictory demands of U.S. foreign policy toward Cuba and make some of their services unavailable on the island. Consequently, the challenges arising from the platformization of the island's emerging networked public sphere arise not primarily from capitalist dominance but rather from U.S. governmental strategies, which officially proclaim to promote the opening of the country.

Even though Cubans are what Jenna Burrell (2011, 142ff.) has called their "unimagined users," they have still harnessed these profit-oriented platforms for innovative purposes. They have utilized them to level the playing field for access to the public sphere, for example, by using Facebook livestreams to express unfiltered grievances, sharing memes that mock a hypernormal-ist state discourse disconnected from reality, or deploying YouTube videos to present the views and opinions of everyday people. The emergence of a networked public sphere that the government cannot contain also paved the way for new forms of independent social organization, which ultimately led to the events of July 11. However, many of the online support networks that activists relied on to organize protests and advocate for change have faltered due to state repression and the resulting mass migration of hopeful young Cubans who have lost faith in the prospects of improvement. This unprece-dented exodus not only drained the opposition of its strength but also deeply impacted the private sector, posing a threat to the country's economic future. Several Cuban university professors told me how their classes were getting emptier by the week as their students disappeared one by one. Even in Havana's vibrant neighborhoods, restaurant owners faced challenges in finding waitstaff, a role once popular among young Cubans due to the potential for tips. The connected Cubans depicted in this chapter are likely to thrive wherever they find themselves. Whether their homeland in the long run can thrive without them is a different question.

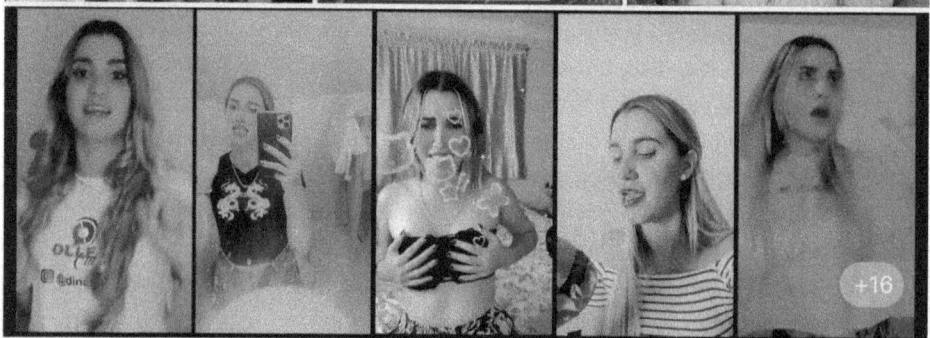

DOCUMENTAL sobre MI VIDA en ALEMANIA / Vlog en el evento y más... 🤍▬

Dinamita (2022)

Ethnographic documentary, HD, 25:25
BY STEFFEN KÖHN AND PAOLA CALVO

www.da.gd/millenials

Starring:
DinaStars
ComePizza
Jhans Oscar

Cinematography: Paola Calvo
Editing: Ginés Olivares
Sound design: Jan Pasemann
Music: Johannes Klingebiel
Animation: Alexander Bley
Color grading: Claudia Gittel
Postproduction: House D-Facto Motion Berlin
Postproduction coordination: Jessica Jakobs
Graphic design: Johannes Büttner
Producer: Paola Calvo
A TUMULT production

FIGURE 5.2. Screenshots from *Dinamita* and DinaStars's YouTube video about the film's world premiere.

Dinamita is an ethnographic documentary that Paola Calvo, a Venezuela-born camerawoman and director, and I filmed between 2018 and 2021. The film revolves around Dina (DinaStars) and Adriano (ComePizza), who are among the pioneering YouTubers in Cuba. They were part of a vibrant community of young video bloggers who used to hang out together on weekends in Calle G, located in Havana's Vedado district, which has long been a hub for the city's subcultures. What made Dina and Adriano stand out was the distinctive content they produced. Adriano performed stand-up comedy with biting commentary on the daily challenges of life in Cuba, while Dina shifted her content from makeup tutorials to discussing women's rights issues after experiencing a sexual assault. We were immediately captivated by their enthusiasm to make their voices heard in Cuba's emergent networked public sphere and the creativity with which they navigated technological, material, and political limitations. Video blogging became their way of carving out a space for themselves, expressing their individuality in a country where public discourse is limited and state-run media rarely represent youth culture.

During the filming process, we adopted a highly performative approach to capture the everyday experiences of our young protagonists. Instead of conducting traditional interviews, we encouraged Dina, Adriano, and their friends to engage in spontaneous dialogues with each other, allowing them to share their thoughts and experiences organically. To create a visual contrast to the fast-paced aesthetics of YouTube, Paola and I decided to work with long sequence shots that are rarely seen in documentary film. We carefully coordinated the camera's movements with the protagonists' own movements in physical space. In pre–mobile Internet Havana, where people would go "to" the Internet rather than "on" it, Dina, Adriano, and the others were always on the move through public spaces. We wanted the film's visual style to reflect this sense of drifting, employing a gliding camera that remained in constant motion. For maximum freedom of movement, we utilized a gimbal and attached lavalier microphones to our participants' bodies. Although we aimed to film each scene as a single take, we allowed ourselves some editing flexibility during postproduction.

In addition to filming original footage, we extensively collected videos from our protagonists' YouTube channels and Instagram reels, where they often corresponded with each other. This material, showcasing the unique aesthetics of Cuban Internet culture, became a crucial layer in the film, enabling us to tell their stories in their own words and from their own perspectives. Over the course of three years, we witnessed significant transformations in their lives. Adriano and particularly Dina rose to fame as internationally recognized influencers. Eventually, Dina's involvement in the July 11 protest prompted her to leave the country, following Adriano, who had already emigrated to the United States a few months earlier.

Working with protagonists who were highly self-conscious about their image and adept at audiovisual storytelling was a fascinating process. At times, Adriano and Dina even produced video clips specifically for our film, helping to explain the intricacies of Cuban Internet access to the audience. On various occasions, we found ourselves becoming the subjects of their videos as well. For instance, when we invited Dina to the film's world premiere at the Hof International Film Festival, she documented her journey to Germany through a series of stories and videos. Our film thus became part of a transmedia dialogue between us, our protagonists, and their followers, several of whom wrote to us to see the film. Consequently, our film not only documented the Cuban social media sphere but actively participated in it, finding an audience beyond conventional platforms such as documentary film festivals and academic conferences.

The QR code provides access to a screener file of the film, stills, the poster, as well as Dina's YouTube video and Instagram stories from the world premiere of the film—an event I regrettably missed as I was in Cuba conducting another period of fieldwork.

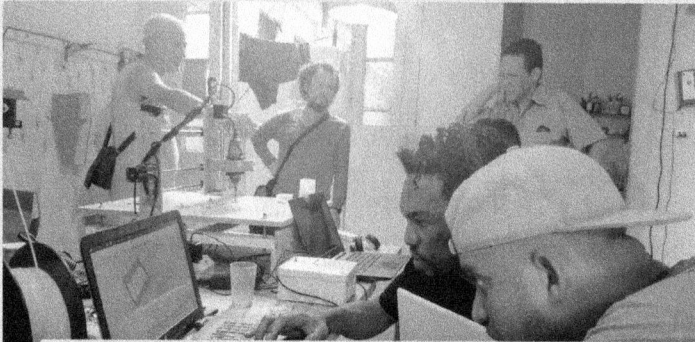

EL LIBRO DE LA FAMILIA

CON NUESTROS PROPIOS ESFUERZOS

CAFÉ' SORPRESA

@copincha
https://t.me/copincha

EL SOLARPUNK

[como ficción
y modelo
de futuro]

Viernes
30 abril
5:00 PM

copincha

BIOHACKING
in pursuit of
human rights

Michael S. Laufer
FOUR THIEVES VINEGARD

FOTOVOLTAICA
EN CASA

copincha

Viernes 6
diciembre
5:00 PM

Laboratorio de
Investigaciones
Fotovoltaicas

Ánimas 564
Apto. 101
entra Soledad
y Oquendo
Centrohabana

CHAPTER 6

Repair

Upon my return to Havana at the end of 2020, when Cuba reopened its borders after a challenging summer grappling with the COVID-19 pandemic, my days were filled with reconnecting with both new and familiar friends and acquaintances. One of the first people I went to see was Erick, the administrator of the SNET backbone. As I ascended the stairs of the socialist housing block in which he lived, I could already hear him eagerly awaiting my arrival in the stairwell. His broad grin revealed his eagerness to share something with me. "Look," he exclaimed, running his hand over a small white box positioned to the right of the entrance of his apartment. With a resounding click, the door unlocked. I stood there, struck with a mix of astonishment and disbelief. Erick chuckled, visibly pleased with his nerdy party trick. "You missed a truly extraordinary event," he revealed. Prior to our meeting, I had already heard about Erick's involvement with Copincha, Havana's pioneering hacker and makerspace, located just a few streets away, not far from Parque Trillo, in the house of Mauri, a designer, doer, and thinker with big ideas and grand aspirations. "We had a visit from this eccentric American biohacker, and the evening culminated in an implant party," Erick recounted. He extended his hand to reveal the back of it, spreading his thumb and index finger apart. "Now, right here, underneath my skin, I have an RFID chip that opens the door. Pretty cyberpunk, don't you think?"

In this final chapter, I delve into a remarkable group of passionate tech enthusiasts who unite against the prevailing sense of stagnation and despair that has permeated the country due to ongoing economic woes and political stalemate. Their collective commitment lies in forging distinctly Cuban solutions to both

FIGURE 6.1. The Copincha hackerspace in Centro Havana offers a range of workshops on topics such as plastic recycling, 3D printing, and repairing technological devices. The collective develops ingenious designs, rooted in the Cuban resolver ethos—a mindset embodied in the self-help books published by the government during the economic crisis of the 1990s. Many of their workshops further draw on cyberpunk and solarpunk ideas.

local and global problems, while nurturing a vision of a self-governed and self-moderated community that propels societal advancement through the free and open-source utilization of technology. I explore the ways in which members of the Copincha community have started to broaden their perspectives beyond the limitations of the island, cultivating transnational connections with like-minded collectives around the world. They actively engage in global social movements that tackle urgent issues such as sustainability, collaborative modes of consumption and production, and the pursuit of sustainable, long-term design. Copincheros strive to contribute to the global quest for alternatives to capitalist extractivism and the wastefulness of planned obsolescence, drawing inspiration from Cuba's unique experience of enduring decades of material scarcity and its profound tradition of repair culture. The practices and concepts they bring to the debate are extrapolated from the local, community-based solutions that Cubans have developed in response to economic and political limitations, such as the decentralized networks and people-embedded infrastructures described in this book. However, while many of their fellow citizens engaged in make-do-and-mend practices out of necessity and gave little thought to the broader potential of their inventions, Copincha members viewed them as much more than mere outcomes of a life of relentless scarcity. Instead, they positioned the Cuban resolver ethos as a future-focused attitude that could offer much-needed resilience in the face of looming global environmental disasters and serve as blueprint for deliberately low-tech approaches that combine resource efficiency, decentralization, and ecological awareness.

In what follows, I show how the group has practically implemented DIY and DIT ("do it together") philosophies, focusing on open-source methods, knowledge sharing, and the goal of developing local answers to what increasingly become global challenges. Copincha members work on projects on a wide range of topics, including permaculture, plastics recycling, and 3D printing. They invest particular energy in carefully documenting their transdisciplinary and sustainable practices so that they can be replicated by others. They also seek to support other groups by providing space and access to knowledge to build a network between kindred initiatives.

The Beginnings of Copincha

Mauri, the visionary behind this collective, emerged as a prominent figure within the tech communities of Havana that I frequented. Following his graduation as an industrial designer in 2013, he swiftly grew disillusioned with his position at a state-owned company specializing in forklift manu-facturing. Since most of the components needed had to be imported from China and only the chassis was made in Cuba, and the company followed a rigid, centralized government plan, there wasn't much room for custom-izing production to meet specific local needs. This experience instilled in him a desire to work on projects that solved actual problems and took into account the perspectives of the people he was designing for. When he spent some time in Lancaster, where his then-partner was studying, he was first exposed to the maker and fab lab culture and began to par-ticipate in a local repair café. He was particularly drawn to the Arduino open-source electronics platform and the worldwide community of makers that has gathered around it. Arduino boards are microcontrollers that can be used for all kinds of physical computing projects, such as interactive systems that use sensors as inputs to control outputs like lights, displays, or motors. When he stumbled upon a handbook by Arduino guru Simon Monk, he wrote him an email to ask if there was a Spanish edition of his book. To his surprise, Monk replied immediately, sending him several PDFs in Spanish and suggesting they meet at Lancaster University since he lived nearby. During their meeting, Monk told Mauri about his plan to return to Cuba, where he had spent his honeymoon years ago, to teach for a week and vacation for a week. Mauri, in whom the idea of importing the makerspace idea to Cuba had been maturing for some time, immediately agreed to organize the workshop and made his apartment in Centro Ha-bana available.

So, a few months later, about fifteen Arduino enthusiasts gathered in Mauri's living room to listen to Simon Monk and work on their projects under his guidance. The creative energy released by this workshop inspired Mauri to permanently donate a floor of his apartment to the community, and thus Copincha was born. Mauri describes Copincha as a community laboratory. *Pincha* means work in Cuban slang, so Copincha literally means collective work. Since the inaugural Ardunio workshop, it has attracted a diverse group of people, from engineers and scientists to designers and art-ists to hackers and hobbyists. Together, they organize talks, workshops, and informal meetings where they discuss the social impact and potential of various technologies.

A recurring event at Copincha is the community-organized talks that draw a wider audience of interested guests. Copincheros have experimented a lot with the format of these lectures and found that these events are particularly suited for bringing new people on board and for exploring topics that could then inspire smaller groups of people to go deeper and organize a laboratory or workshop around the issue. The three ongoing series of talks are Café Sorpresa (Café surprise), Café Reparación (Café repair), and Café Viaje en Casa (Café travel at home). The idea of Café Sorpresa is that anyone who knows about a topic that might be of interest to the community is invited to share their experiences. In its deliberate negation of hierarchies between different forms of knowledge, the Café Sorpresa program tries to maintain an element of surprise, so that the audience always finds something strange, new that they did not expect. For example, a speaker could be someone a Copincha member met on the street and invited to talk about how they repair shoes or just as easily a university professor who was working on an interesting aspect of technology. The biohacking workshop that ended with Erick and seven other Copincha members getting RFID implants also began as a Café Sorpresa.

Café Reparación, on the other hand, is very much connected to practices of repair that are an important element of global maker culture, especially in the Ifixit movement or the so-called Restart parties that Mauri had encountered in the United Kingdom, where volunteers help people fix their broken devices. Café Reparación usually starts from a concrete problem that someone in the group has, a broken laptop hinge for example, and develops a collective activity from it. The event is also intended to open up a space for reflection on the owner's relationship to the object as well as to demonstrate the potential of a collective approach to problem solving. As I have discussed in previous chapters, the ability to repair things is an essential part of Cuban resolver culture and a necessity for coping with the island's persistent shortage of materials. At the meetings I attended, however, copincheros were less interested in discussing repair in terms of scarcity and instead invoked global discourses of sustainability and resistance to purposeful obsolescence. Participants were familiar with international attempts to introduce a legal "right to repair" and knew exactly which tech companies were most restricting access to tools and parts, or even setting up software barriers to prevent independent repair.

In a third series of presentations, Viaje en Casa, invited speakers share their experiences of traveling to another country. The talks are usually accompanied by photos, and the presenters offer many practical tips for getting along in the country in question. Yet I felt that for most of the participants, many of whom had never left the island, Viaje en Casa primarily opened an imaginary space that allowed participants to see themselves as global subjects. I was often surprised by how detailed the questions from the audience were, especially from people whom I would not have expected to have the desire (or the means) to visit the country in question. International travel is still an almost unattainable privilege for most Cubans, and those who have visited other countries usually did so because they had family there or had worked as mulas, not because they freely followed their curiosity and interests. Amid the escalating wave of migration following July 11, accompanied by subsequent political efforts, particularly from the United States, to impose restrictions, the possibility of international travel for Cubans was in danger of reverting to its preliberalization state, when it primarily served as a conduit for permanent emigration rather than facilitating temporary visits. Given the country's bleak economic outlook, I sensed that there was something cathartic about the engaging discussions about the peculiarities of foreign cultures and the collective fantasy of an international lifestyle that Viaje en Casa allowed.

Just like the lectures, the workshops at Copincha are also conducted according to open-source principles. This means that every participant should master the practice in question afterward and ideally be able to pass on their knowledge. A prime example of this approach is Nestor's project *CubaCreativa [GUARAPERA]*, for which he crowdsourced local knowledge to develop the perfect Trapiche, a machine for extracting sugarcane juice. Selling sugarcane juice is a common side job for many people on the island, and since this equipment is not commercially available anywhere, most vendors have had to build their own machines to produce the juice. After interviewing dozens of vendors and months of prototyping at Copincha, Nestor and his collaborators created a technical manual for building an optimized sugarcane mill that combines the advantages and features of all the local designs they studied and is based on materials that are more readily available in Cuba. This free manual, created for sharing, includes details on the design of the machine, a list of all the components required, step-by-step construction instructions, and information on how to adapt it to different production volumes. It also discusses the use of alternative energy sources and the possible uses of the waste produced during juice production.

With the assistance of Luis Rodil-Fernández, a devoted supporter and collaborator of Copincha from Spain, who also serves as a professor of interaction design at a university in the Netherlands, the group received the necessary components to construct a 3D printer. Since then, a significant portion of the group's recent endeavors has been centered around plastic recycling. While plastic waste is a serious problem in Cuba, recycling also gave the group access to the raw material needed to use the printer. Their initiative even received a grant that allowed them to use the facilities of Habana Espacios Creativos, a brand-new creative space in Habana Vieja funded by the European Union and managed by the Oficina del Historiador de la Habana (Office of the Havana Historian).[1] There, Mauri and his partner José Luis, who have been instrumental in driving the project forward, regularly organized hands-on workshops to get more citizens involved. In these meetings, they went through the entire recycling process with participants, from sorting and cleaning the plastic to using the shredding machines to melting and pressing it into filament for the 3D printers. Copincheros had built all the machines for processing the plastic themselves from recycled or reused parts of other equipment, as well as from components and donations from international groups. Assembling the machines was a time-consuming and tedious process. It took Mauri a full year to find someone capable of building the continuous screw for the extruder machine that produced the filament, but he came to accept this as an inevitable part of expanding the Copincha network:

> I finally came across someone who had the necessary tools and decided to give it a shot. Once he successfully made the screw, there were several people who expressed their interest in obtaining one as well, after I shared the news of his accomplishment in our Telegram group. So despite the slow pace of building these machines, we started forming a network of people who were connected to us, contributing their skills and resources. Even though the process initially began slowly, the activated connections grew stronger over time, allowing us to progress more rapidly.

Mauri and José Luis also approached other local initiatives involved in recycling plastics as well as private businesses (like the popular independent fashion store Clandestina) and government institutions (such as the University of Havana's Faculty of Psychology) to create a citywide network of plastic collection points. Among the first objects they produced in larger quantities were face shields to protect doctors and nurses during the coronavirus pandemic, cell phone holders for bicycles, and wheels for crates. They made their designs available as STL files to anyone interested and also experimented with open-source printing models provided by Precious Plastic, a global open hardware plastic recycling community of which Copincha eventually became a local node.

Building (and Maintaining) a Community

During one of the first meetings that took place after Simon Monk's Arduino workshop, the budding community convened in Mauri's living room with a shared objective of delving into the realm of home automation. However, since they lacked access to any domotics hardware, the only tangible result of the reunion was the drilling of a hole in the wall that separated Mauri's living room and bathroom. This seemingly mundane act of drilling eventually became the foundational story of Copincha. While Mauri had already distributed keys to the place to several group members, this pivotal moment became a significant turning point for him, as the hole transformed his living room into a truly shared community space. It symbolized the collective ownership and autonomy of the members, who now felt empowered to make modifications without seeking Mauri's permission as the sole owner of the place.

Copincha's physical space cannot accommodate more than about a dozen people at once. Yet because Mauri has sought from the beginning to build connections with other groups and as many members also have their own workspaces, Copincha's network of collaborators has been constantly expanding. The grant for the plastic recycling project from Habana Espacios Creativos eventually allowed the initiative to temporarily occupy a larger space and hold events with much more people, and on Telegram Copincha has also brought together a lively online community. Copincha membership is therefore somewhat fluid and allows for different degrees of involvement. While *la familia* (the family) comprises the people who are permanently committed to the group and its physical space (*la casa*), *la comunidad* (the community) also includes people who are only loosely connected to some of Copincha's activities.

The family meets regularly and invests considerable time and effort in developing community organizing tools and a culture of shared decision-making. A key decision was that each family member would be expected to contribute to the upkeep of the space and the collective. Over the years, the group has agreed on different forms of contributions, such as a monthly monetary donation of 2 CUC (1 CUC for students and those who cannot afford the regular amount) or a nonmonetary contribution in the form of things needed to run the workshops and organize activities. Members can also give their free time by, for example, sharing their knowledge, giving classes or workshops, and cleaning the space. Decisions about how to spend the funds (e.g., for buying tools or machine parts) and what projects the community would like to pursue are taken at the weekly Sunday meetings. All decisions are made by consensus, and each member has access to the spreadsheet that documents all income and expenses.

Mauri, who, because he provides the physical space for many of the group's activities, is inevitably the primus inter pares, described Copincha's horizontal and outward-looking form of organization as a "permacultural" way of building relationships inspired by the *conuco* cultivation system of Indigenous Taíno peoples in the Caribbean. Conucos were small community gardens near Taíno villages for which plots of land were cleared and burned and then planted with staples such as root crops.[2] Because these crops matured at different rates, nearly continuous harvests were possible. Like a conuco garden that yields a bountiful harvest year-round, Mauri envisioned Copincha as a space to cultivate knowledge, practices, and ideas that nourish the community.

In addition to the conuco, he conjured up a series of concepts to bring to life the vision of collaboration that Copincha was striving for. For him, the process of building a community began with *conocer* (knowing), which referred to the specific knowledge or interest each new member brought to Copincha. The next step was *conectar* (connecting) one's knowledge with that of other people. Resources and ideas were collected with the aim of *construir* (building) something. Everything created with the community's resources should belong to the community and serve it first and foremost. This ethos of *colectivizar* (collectivizing) was meant to help Copincha achieve greater self-sustainability and autonomy. *Comprometer* (committing) then referred to the process of taking up responsibilities within the group, while *coexistir* (coexisting) was about reflecting the collective's outward-looking relationships with other communities.

Such sophisticated ideas about forming an independent and self-sufficient community unsurprisingly led to a complicated relationship with the chronically "autonomy-phobic" Cuban state institutions. Since the members of Copincha come from all parts of Cuban society, there were also some among them who worked in the military or the police and who therefore could not participate, for example, in talks given by foreign guests. Copincheros tried to respect these boundaries and resolve such dilemmas by making recordings of the events available to interested people who could not attend for political reasons.

Much of what happened in Copincha was relatable to broad segments of the population, as building and repairing things are crucial skills on the island. At the same time, many of the group's ideas about sharing and collectivizing knowledge as well as about volunteering for the community are very closely related to the progressive thinking espoused by the Cuban Revolution. During the Special Period, when the state needed to inculcate resilience and survival in its citizens, the Cuban military published two books that are still a reference for copincheros. *El libro de la familia* (The family book) and *Con nuestros propios esfuerzos* (Through our own efforts) were both edited by an anonymous collective of authors and gathered essential knowledge for a self-sufficient life with minimal resource consumption and without depending on imported goods or energy. They sought to teach Cubans, among other things, to grow their own vegetables, substitute fertilizers, breed animals, make their own cleaning products, produce herbal medicine, and build their own tools, furniture, and machines. The second book, *Con nuestros propios esfuerzos*, in particular, documented the various survival strategies that people all over the island had found to withstand the difficulties of these new times. They were collected by the local mass organizations on the ground to share with the entire population the different initiatives, experiences, and efforts that were blossoming everywhere in the country.

In spite of these ideological agreements, it was difficult for state institutions such as universities to cooperate with Copincha because as a private initiative it had no legal form in which the state could have recognized it. As Mauri explained, "That takes away a lot of possibilities: For example, whenever we want to collaborate with the university, the university has to follow this legal framework and asks: 'What is Copincha, what institution is Copincha?' We can't just work with them, even if we have all the knowledge, the relationships, and the resources." Somewhat ironically, given the shared ideals, the form in which the state was most ready to accept Copincha was not that of a civil society initiative or NGO but that of a business. Mauri therefore explored ways to register Copincha as a cooperative while maintaining its social purpose. This tension between a business approach and the ideology of sharing had to be constantly renegotiated within the community as well. Especially in the beginning, new members who earned their money in the informal sector were often skeptical about sharing their knowledge (which often gave them a competitive advantage in the free market) in the talks and workshops, which required a lot of convincing by the Copincha family. Some of the things designed in the workshops were also developed into marketable products by members of the community, so that over time several small businesses were created around Copincha, selling, for example, hydraulic fittings made from PET bottles or bags from recycled plastic.

The Copincha family tried to face these contradictions between the logic of sharing and the logic of the market not with ideological rigidity but with flexibility, knowing that the development of marketable products helped them achieve legal recognition in the country (especially after the state took further steps to create a regulatory framework for small and medium enterprises after the events of July 11).[3] Accepting that some of the results of their work would be turned into products, they began to define themselves primarily as a learning and knowledge-sharing community, as this was the best way to realize their ideals of horizontal distribution, open source, and self-reliance.

Creating a Digital Infrastructure

Digital communication infrastructure played a vital part in holding the community together. Mauri remembered how before mobile Internet they often had to spend considerable sums of money to send SMS messages to all Copincha members to inform them about events. After the rollout of the nationwide 3G network, Telegram quickly became the group's main communication tool because of the free cloud space it provides and its videoconferencing function. Members began to use Copincha's Telegram channel, for example, to continue the discussion after the lectures or as a barter platform (see chapter 4). To separate the different uses, they opened several thematic channels for different topics each, for example, an archive channel for uploading recorded presentations and a repository of pirated PDF books on topics the community was working on, such as permaculture, fermentation, and sustainable design. Moderators of these channels made a lot of use of Telegram's feature of editing posted messages and creating links between messages to better structure the flow of information and make them function more like a website. During the COVID-19 lockdowns, Telegram quickly became the central community hub, where the regular program of talks continued via audio messages.

Copincha's projects have often been constrained not only by the limits set by the Cuban state but also by the impact of the U.S. embargo on Cubans' access to digital tools. This became once more apparent when Mauri wanted to add all the plastic recollection points that had been set up in Havana to the global community map on the Precious Plastic website. He found that this was impossible as these data were stored on Google's cloud database Firebase, which is restricted for Cuba. So although Precious Plastic was using a decidedly participatory crowdsourcing tool in the form of an interactive map that anyone could contribute to, it still excluded Cuban users, which meant copincheros had to build their own map using open-source software.

Experiences like these motivated the group to apply their ideas of independence and self-reliance to the digital realm, seeking alternative tools that could mitigate the restricting effects of U.S. digital imperialism and be easily adapted to the characteristics of the Cuban digital environment. This growing desire for technological sovereignty became especially important in a project that began to occupy copincheros more and more. The realization that many citizen initiatives in Cuba were unable to achieve greater social impact because they lacked appropriate communication and documentation tools made the group aware of the need to document the knowledge that members had collectively acquired and make it available for others.

Copincha members eventually found the instruments they were looking for through their participation in DOTS—The Impact Summit (an online congress that brought together innovation centers and tech communities from around the world, organized by the Berlin-based NGO Global Innovation Gathering), when they connected with initiatives that developed software solutions for and from the Global South. At the event, Offray Luna Cárdenas, one of the cofounders of the Colombian hackerspace HackBo, presented a set of self-developed digital tools for collaboration that he called "pocket infrastructures." These are modifiable, flexible, and simplified infrastructures designed to amplify participation (Luna Cárdenas 2019). Because pocket infrastructures do not depend on third-party providers, work well in contexts where connectivity is low, expensive, or nonexistent, are compatible with most operating systems, and can be run from a variety of hardware (such as, for example, a USB stick, a Raspberry Pi single-board computer, or a modest server), these tools were perfectly suited to the Cuban reality.

Copincheros began prototyping Copinchapedia, an open archive in which they documented their collected designs, experiences, and reflections with a range of pocket infrastructure software. They used Grafoscopio, a tool for interactive documentation and data visualization that enables the combination of text, data, code, and visualizations into interactive, reproducible documents, the version control system Fossil that allows coordinated work on a shared file, and TiddlyWiki, a wiki application that can be used as a collaborative, nonlinear notebook for organizing and sharing complex information. What all these tools have in common is that they are self-contained, meaning that they wrap all data together in one executable file. TiddlyWiki, for example, saves the wiki and the program itself in the same html file, making it readable on any computer without preinstalled software. Therefore, these applications are ideal for contexts like Cuba where people often work on old machines with unlicensed and/or unupdated software and online data transfer can be cumbersome, but where sophisticated offline infrastructures for peer-to-peer distribution are in place.

To make Copinchapedia accessible to a wide audience, copincheros found el paquete semanal to be the perfect medium to fit their philosophy of autonomy, self-sustainability, and bottom-up citizen innovation. The data generated with pocket infrastructures also fit well with the demands of the paquete matrices that always ask Cuban content producers to reduce their material to as few files as possible to prevent loss of content during the copy process. Copincha members thus began to use this decidedly local infrastructure to share their projects with their fellow citizens right at a time when Cubans finally had more unrestricted access to global commercial social media platforms and Facebook and WhatsApp had become important spaces for public debate (as detailed in chapter 5). While criticism of the hegemonic power of these digital behemoths over the information and interpersonal communication of their users had barely penetrated the consciousness of broader Cuban society, this was a much-discussed topic at Copincha meetings. Like all of the grassroots tech initiatives from both the Global North and the Global South to which they were connected, they were concerned that profit-motivated platformization was eroding the potential of digital tools for commons-based peer production as well as the foundational promise of the open web. Using open-source software and relying on the preexisting interpersonal circulation networks that had developed in Cuba was not only the most practical solution for these young designers, engineers, and artists, but also a "proof of concept" demonstrating that it was possible to share knowledge without depending on global tech monopolies with their exploitative and extractive practices. It also allowed them to reject the top-down communication of the Cuban state, which copincheros saw embodied, for example, in the ideological tone of survival manuals published by the Cuban military during the Special Period.

Transnational Alliances, Global Ideas

As I have emphasized throughout this chapter, Copincha is embedded in a transnational network of collaborators with whom members share ideas and from whom they occasionally receive help. These international allies were mostly people from the global maker community, environmental activists, or sustainability and low-tech advocates who had learned about Copincha online, made contact, and eventually came to give talks or workshops. After the relaxation of travel restrictions under Obama, a steady stream of American makers, hackers, innovators, and roboticists came to Copincha to present their work and often left equipment and materials for the community. While he was pleased that his projects resonated with like-minded people in other parts of the world who were actively seeking alternatives to the monopolization of technology, environmental degradation, and resource waste under capitalism, Mauri often expressed frustration with the perceived lack of interest in these issues among many of his fellow Cubans, for whom practices of repair, reuse, and recycling implied poverty, not innovation. He maintained that the historical experience of the economic downturn after the fall of the socialist camp had led Cubans to see repair as an outcome of material scarcity and not as a future-facing practice. As discussed in chapter 3, I encountered the widespread longing for smoothly functioning and always available technology, for example, among many SNET users who would trade their fragile homegrown network in a heartbeat for the universal and affordable access to the Internet available for most citizens in the Global North. My friend Yasser's bicycle activism (chapter 5) encountered similar backlash, as many of his compatriots still associate bicycling with the hardships of the Special Period, when gasoline shortages forced the population to get around using their own muscle power.

Copincha established connections predominantly with groups and individuals from Europe or the United States (and rarely with other communities from Latin America) not only because of similar ideas and worldviews but also because there is simply much more funding for NGOs and civil society initiatives available in these countries. Even the group's productive collaboration with the Colombian hackerspace HackBo came about only because they met through an event organized by a German NGO. Offray, aware of the asymmetry of resources such as funding and government support that characterize such "South-North-South" triangular relationships, pointed out that he has met more people from other Latin American countries doing similar things to him at events in Europe than at meetings in his hometown of Bogotá. Although for these reasons Copincha's growing global network tends to extend to the Global North (rather than to the many Latin American, African, and Asian countries that have historically maintained strong diplomatic and cultural ties with Cuba), it provides the community with a constant inflow of new guests and collaborators and thus a connection to the dominant issues and discourses being discussed in these parts of the globe.

During one of my conversations with Mauri about the profound impact of this ongoing transnational intellectual exchange on Copincha, he humorously expressed that copincheros do not consider themselves as citizens of the world but rather aspire to be *ciudadanos del mundo roto* (citizens of a broken world). Echoing information science scholar Steven J. Jackson's (2014) influential call for "broken world thinking," which takes erosion, breakdown, and decay as starting points in thinking through the social, economic, and technological systems we inhabit, he asserted,

> Our Cafés Sorpresas are about looking at repair, recycling or reuse capabilities as the skills of the future. In a future where many of our infrastructures may no longer exist and resources are diminishing, such skills will be in demand. So becoming a citizen of a broken world means preparing for that possible future, and in that sense Cuba might even represent a future for the rest of the planet, because in Cuba we already live in a postapocalyptic world.

Viewing Cuba—a country already impacted by extreme weather events and rising sea levels due to climate change, and whose citizens have endured a complete economic meltdown and defied material austerity for decades—as a laboratory of the future struck a chord with numerous international visitors to Copincha. Many of them were in search of a blueprint for life after extractivist capitalism, recognizing the potential lessons to be learned from Cuba's unique experiences. For example, when Copincha's longtime ally Luis introduced Mauri to students during a workshop they held together at a Canadian university, he described him as "someone who comes from the future, from our common future. A future where we have to work as a community and without fossil fuels."

Such speculative thinking was an important aspect of the practice of Copincha, whose members shared as much science fiction literature as permaculture manuals on Telegram, indulged in collective imagination during the Viaje en Casa events, and frequently invited Cuban sci-fi authors to give talks or presentations. As Mauri stated,

> I've always enjoyed reading about utopian or dystopian worlds to think about our future here in Cuba. At some point I saw projections that this part of Centro Habana, where Copincha is located, could be covered by water in twenty or thirty years. That's why we built all the furniture here with wood I found on the street, or from the boxes of forklift parts from my workplace. People said, "What are you doing? The wood could be infested with termites," but I wanted furniture that could float on water. At that time, I had a very strong vision of these dystopian worlds of science fiction.

From its inception, the Copincha family has embraced such speculative thinking as a core principle. Simon Monk, the patron of the community, frequently referenced his acclaimed book, *The Maker's Guide to the Zombie Apocalypse* (2016), during his workshops. The book provides practical instructions for using simple circuits, Arduinos, and Raspberry Pis to generate electricity, reuse materials, create essential electronics, and build a communications infrastructure for surviving a return of the dead (and other postapocalyptic scenarios). Naturally, this also drew the interest of Havana's vibrant community of science fiction writers, who explored Cuba's possible futures within their literary works. In their writings, these authors delved into concepts akin to those explored by copincheros. Many of them specifically focused on the increasingly dire consequences of climate change on the island, crafting narratives featuring street-smart individuals who adeptly appropriated technology, resisting repressive institutions while navigating alternative social realms within cyberspace. Cuban cyberpunk, the island's dominating sci-fi subgenre, is a literary movement that has indigenized the cyberpunk literature from its northern neighbor to reflect on the country's technological and social reality, as literary scholar Juan Toledano Redondo (2005, 2019) has discussed. Cyberpunk novels from the United States began trickling into Cuba in the 1990s through the informal media distribution networks described in chapter 2, at the very moment when after the collapse of the Soviet Union the Cuban economy went into free fall.

U.S. cyberpunk is set in a near-future dystopian urban environment of oppressive transnational capitalism, environmental deterioration, and rapidly evolving technology that has invaded the human body. It created a highly influential iconography for the digital age before the Internet became ubiquitous. Cuban cyberpunk has taken up the genre's tropes of social decay to make sense of the country's particular geopolitical situation, seeking to escape, as Cuban author Vladimir Hernández (quoted in Redondo 2005, 449) maintains, from the superficial local provincialism that has characterized the work of previous generations of sci-fi writers. The genre import allowed them to break with older literary models and their ideological messages and positive socialist heroes. The protagonists of Cuban cyberpunk literature are often uprooted, trying to survive in a future of economic ruin and maladjustment to a brutal capitalism. Frequently, ecological disaster has already occurred, as in Erick J. Mota's *Habana Underguater* (2010), where half the city has sunk after a massive hurricane, residents are engaged in a civil war in physical reality and cyberspace, and hackers are reassembling technological devices to maintain fragile and ad hoc computer networks.

Mota's work in particular resonated deeply with copincheros, and his regular exchanges with the community fueled a vibrant ecosystem of speculative minds. This convergence of visionary thinkers within Copincha sparked innovative ideas and invigorated discussions about the potential paths for Cuba's society. As Copincha members strongly identified with the protagonists of these narratives and the worlds they inhabited, many talks and workshops conveyed a cyberpunk sensibility that applied the genre's worldview to reality. They invoked a hacker approach to appropriating technology, which meshed well with the Cuban resolver ethos. Michael Laufer of Four Thieves Vinegar, a U.S.-based collective of anarchists and biohackers developing DIY medical technologies, whose talk at Copincha ended in the infamous implant party, was just one of many speakers who demonstrated how "the street finds its own uses for things" (as William Gibson [1982, 106] famously noted in his genre-defining short story "Burning Chrome").

While cyberpunk allowed them to think about personal agency, enabling individuals to overcome larger, impersonal forces that control their lives, becoming a "citizen of a broken world" for Copincha members also meant building durable community structures and not settling for a dystopian status quo. Thus, they also drew inspiration from cyberpunk's twin genre, solarpunk, to explore less pessimistic alternate futures. Solarpunk is a literary, aesthetic, and activist movement that has gained prominence over the past decade and seeks to create utopian counterimages to cyberpunk's nihilistic dystopia. It shares with cyberpunk its technophilia and countercultural, anticonsumerist, and egalitarian "punk" ethos but rejects its defeatist view of societal future and instead envisions positive ecological and social change. Proponents of this Internet-born genre that first flourished on platforms like Tumblr, Reddit, and Pinterest seek above all to visualize a desirable future in which humans have overcome their dependence on fossil fuels, balanced nature and technology, and explore new ways of living together.

Even though its first advocates came mainly from the United States, solarpunk has evolved into a truly global movement. One of the very first anthologies of the genre was published in Brazil (Lodi-Ribeiro 2012), and the genre celebrates practices of technology repurposing, decentralized networks, and DIY solutions originating from the Global South. As a result, many activist collectives from these parts of the world have embraced solarpunk's aspirational prefigurative politics, creating a cycle of inspiration that has come full circle. These collectives now actively strive to embody solarpunk's values of ingenuity, generativity, independence, and community in their real-life endeavors. HackBo cofounder Offray, for example, who considers himself a solarpunk, told me that he sees the group's communal development and use of grassroots pocket infrastructures as a means to enact alternative futures in the here and now.

In his Café Sorpresa talk about solarpunk as fiction and model for the future, Erick Mota emphasized that the idea of a solar-powered future comes naturally to a country with an abundance of sunshine that has already experienced several severe energy crises. Cuban power generation is notoriously inefficient due to old power plants and an outdated grid, and households waste much of the heavily subsidized electricity through energy-guzzling appliances such as aged refrigerators. While the Cuban state is still highly dependent on oil and gas imports it increasingly struggles to afford and seems unable to address the transition to a decarbonized future, several groups around Copincha have started to work on bottom-up solutions and engage in experiments of technological resilience that involve the intelligent use and conservation of energy. For example, Jorge Luis and his team of engineers and designers have developed a prototype smart meter that can monitor the energy consumption of each domestic appliance, which would allow households to increase their energy efficiency and lower their electricity bills (ironically, on the day he presented the smart meter during a Café Sorpresa, large parts of Havana were again without power). In order to scale up production of the device, they planned to leverage the legalization of private micro and small enterprises after the July 11 protests by establishing a small company. By doing so, they would gain access to state-owned import companies, allowing them to procure the necessary materials for manufacturing on a larger scale.

Another captivating project was conceived by a group of idealistic former SNET members who, disenchanted by the state's appropriation of their network, decided to join forces with Copincha. They created a solar-driven Wi-Fi antenna capable of withstanding the frequent power outages in the country and serving as a community server for data exchange that could function as a local alternative to the expensive Internet access provided by the corporate state monopolist ETECSA. It consists of a photovoltaic-powered Raspberry Pi with an integrated microSD card module or other mass storage device and a small router that allows people nearby to find or leave files or access an anonymous chat room or forum, creating a wireless network that permits connected users to share information anonymously and locally. The device has a stable Wi-Fi signal range of two hundred to three hundred meters, providing reliable connectivity within this coverage area. The group initially managed to produce only a handful of these antennas as their production and distribution were limited by the fact that solar panels have to be privately imported from abroad.

Yet they successfully adapted this technology, which perfectly embodies the solarpunk values of renewable energy use, decentralization, and community orientation, for Nestor's project *Connectify/Free_Wi-Fi [poesía]*, in which he installed Wi-Fi antennas that broadcast poetry at outdoor locations in Havana and Miami (for detailed information on the device, refer to the specific sub-folder in the digital documentation of this project). The work's monthslong exhibition at several locations in Cuba and South Florida thus served as a successful test run for the design and technology.

Conclusion

Members of the Copincha collective have harnessed the ideas from cyber-punk and solarpunk science fiction to cultivate alternative visions for the future of their country during a period of political disillusionment and a dearth of prospects, where many of their compatriots saw emigration as their sole option. They drew inspiration from the narratives of technological self-sufficiency and hacker resourcefulness found in these speculative fiction genres, reimagining long-standing Cuban traditions of repair and reuse as avant-garde practices suitable for a post-crisis world. What the majority of their fellow citizens perceived as a consequence of poverty, they celebrated as the skills of tomorrow.

In one of our long conversations, Mauri contemplated the wide-ranging implications of creating one's own infrastructures. To him, the concepts of hacking and open source signified that anyone, through their small-scale ac-tions, could contribute to an innovative process and generate alternatives. He pondered how our reliance on infrastructures controlled by others, in which we cannot actively participate, impacts our mobility, communication, access to resources, and interaction with the environment. For him, building a modest infrastructure that caters to oneself and one's community offered an insight into doing things differently, suggesting the potential for transforming larger, more permanent structures as well. The forms of active knowledge exchange practiced at Copincha would thus enable citizens to become each other's in-frastructure, according to his belief.

The vernacular infrastructures that Cubans have fashioned to surmount technological, material, and social limitations, as elucidated in this book, in this sense should be viewed as experiments in autonomous self-governance. They have not only addressed practical challenges but also fostered the emergence of a previously absent culture of collective self-organization, empowering citizens to take more control of their own affairs. Although neither the compilers of the paquete semanal nor the administrators of SNET or the Telegram exchange groups would describe the networks they have created and maintained as spaces for political engagement, these structures inevitably have compelled the connected individuals to establish their own decision-making processes and self-administration methods. In a political, economic, and social system that has become largely dysfunctional and undermines citizen autonomy, these experiences, while they may lie below the threshold of what is usually considered political, may harbor the potential for social change.

DAY -
GRADUAL VIDEO DEGRADATION

+15

INTERLUDE 6
Memoria (2023)

Expanded cinema installation with four video channels (16:37),
a central server, custom hardware and software
**BY STEFFEN KÖHN AND NESTOR SIRÉ IN COLLABORATION
WITH ERICK J. MOTA AND COPINCHA**

www.da.gd/pincha

Starring:
Dayron Villalón Pereda
Chris Gómez Gónzalez
Felix Roman
Benito Guerra
Albany López Suárez
Lynn Cruz
Raudelys Rodríguez Pérez

Screenplay: Erick J. Mota
Cinematography: Nicole Medvecka
Costumes: Raki Fernandez
CGI and VFX: Alexander Bley
Additional VFX: Helman Bejerano Delgado
Editing: Miguel Coyula, Ginés Olivares
Dialogue editing: Malte Audick

FIGURE 6.2. Installation views of *Memoria* from Aksioma | Project Space, Ljubljana (2023) and Hong-Gah Museum, Taipei (2023).

Sound design: Lorenz Fischer, Malte Audick
Rerecording mix: Lorenz Fischer
Music: Johannes Klingebiel
First AD: Francesco Innocenti
Software developer and programming: Eduardo Pujol
Hardware design: Maurice Haedo Sanabria
Props and prostheses: Erick Maza

Memoria is an immersive expanded cinema installation that interweaves two narrative planes: the dystopian vision depicted in William Gibson's "Johnny Mnemonic" (1981), one of the pioneering cyberpunk short stories, and the Cuban physical data distribution network el paquete semanal. Coproduced by and premiered at Aksioma Institute for Contemporary Art in Ljubljana, *Memoria* transposes Gibson's story of a data courier to the contemporary Cuban media landscape, where informal infrastructures of data exchange and distribution thrive. In Gibson's "Johnny Mnemonic," the protagonist is a data trafficker who has undergone cybernetic surgery to implant a data storage system in his brain. This enables him to serve as a human medium for transferring digital information too sensitive to be transmitted over the Internet. He earns a modest living by physically transporting classified data for corporations, criminal organizations, and wealthy individuals. The narrative unfolds as Johnny discovers that the latest data package uploaded to his brain has been stolen from the Yakuza, who now pursues him. With the assistance of Jones, a retired Navy dolphin with cybernetic enhancements and a heroin addiction, Johnny attempts to extract the data from his mind.

The idea for this "documentary remake" that reimagines Johnny Mnemonic as a paquetero emerged from a conversation between Cuban sci-fi writer Erick Mota and me, reflecting on how our technological present is haunted by lost futures. The work explores the parallels between the 1980s cyberpunk literature's vision of a global data network and the contemporary Cuban experience of the Internet. We sought to discover Gibson's signature cyberpunk landscapes within Havana's urban tapestry, transforming the city into a surrogate for the "Sprawl," the urban dystopia described by Gibson as a ruin of the future where technological and scientific advancements coexist with the breakdown of social order. The historic Barrio Chino neighborhood in Havana represents the Asian megacities commonly depicted in cyberpunk narratives. Furthermore, we discovered the ideal setting for the abandoned amusement park where Jones resides in the city's equally abandoned Aquario Nacional. The inclusion of drone footage showcasing a flooded Havana, filmed after the passage of Hurricane Irma along the northern coasts of the island and generously provided to us by the Cuban independent journalism platform *El Toque*, is a nod to Mota's *Habana Underguater*, his trilogy of cyberpunk novels set in a futuristic, postapocalyptic Havana where half of the city is permanently submerged in water.

The work was filmed in 2021, coinciding with the year in which the plot of "Johnny Mnemonic" unfolds and during a pandemic, as envisioned by Gibson in his screenplay for the Hollywood adaptation of his story. Alongside professional actors, some actual paqueteros embody Gibson's cast of fictional characters. Many of the props and cyberimplants worn by the protagonists were designed and created by Copincha member Erick Maza, who is a professional jewelry maker.

Memoria's narrative unfolds across four synchronized screens, merging live-action footage with found footage circulating within Cuba through el paquete semanal. These materials were sourced from *ARCA [Archive]*, an artwork by Nestor Siré and Julia Weist that consists of a 64 TB server housing an entire year's content from el paquete semanal, representing the only extensive archive of this ephemeral phenomenon. To evoke Johnny Mnemonic's memory loss, the installation progressively "dies" over the course of the exhibition. To achieve this effect, copinchero and former SNET administrator Eduardo Pujol developed a unique hardware and software setup. It involves a central server, composed of a computer with a video card featuring four 4K outputs. This server synchronously plays back the four video channels and the the 4.1 audio track. The computer runs a bash script that employs various programs, including FFMPG (a powerful free software for video and audio transcoding) as well as two custom Ruby programs: Datamosh, which deletes segments of the video for subsequent overwriting, and Moshy, which replaces these empty segments with information copied from random other parts of the video. As a result, the visual information in the videos appears to gradually deteriorate and the image progressively degrades with each passing day. However, contrary to this perception, the process actually enriches the video files with their own information, leading to an increase in file size during the copy and overwrite process. Since there is no linear header in the file to convert these bits into pixels, they become visible as artifacts that corrupt the videos. The speed at which the material deteriorates is adjusted to align with the duration of the exhibition. The copy script functions akin to a tumor virus, causing the cells inside the host to grow in a random and uncontrolled manner. Each day, the script generates a new set of video files that amplify the changes from the previous set. Subtle changes that may have been initially imperceptible become more pronounced in subsequent generations, eventually resulting in complete distortion of the information over time as the video devours itself through this constant process of faulty replication. Following Copincha's open-source approach, we have made this custom software freely available on GitHub.

To make the hardware an intrinsic part of the installation, we enlisted Copincha founder Maurice Haedo Sanabria to design a custom case for the server and video card that hangs from the ceiling of the exhibition space as a sculptural object. This case was created using injection molding and 3D printing techniques, utilizing recycled plastic filaments at Copincha. By blending elements of the Cuban context with a work of fiction, we aim to explore alternative visions to the mainstream capitalist consumerist version of the Internet that dominates today's reality, drawing inspiration from both science fiction and the real-world Cuban context. The installation prompts us to contemplate what the Internet could have been or could still become: a more decentralized, net-neutral, and noncommercial space.

The accompanying data collection includes a (single-channel) screener file of the video installation, stills, exhibition views, documentation videos, the software we developed for the video degradation process, and the open-source 3D model of the custom case housing the server.

Conclusion

The story of Cuba's digital awakening provides a unique window into the ways in which global digital technologies become "localized" or appropriated across various cultural and political contexts. By tracing what Cubans experience as "the Internet"—a distinct conglomerate of varying horizons of technical and social possibility—this book has explored how the digital-powered structural transformation of the public sphere has unfolded in markedly different ways on the island, diverging not only from patterns observed in liberal democracies but also from those in other authoritarian contexts. Cubans' wide-ranging adoption of digital media and communication technology since the mid-2010s has catalyzed the transfiguration of the country's previously authoritarian public sphere into a burgeoning networked public sphere. This newly established sphere includes a variety of (semi-)independent social and communicative environments, granting Cubans unmatched political and economic possibilities and profoundly altering the dynamics between the citizenry and the state. It has led to the opening of new communication channels between citizens and government officials, the creation of new horizontal connections that allow citizens to unite over shared topics or concerns, and the reduction of their economic dependence on the state. In concluding, I highlight how the Cuban example enriches our understanding of the interplay between publics and social media platforms, the nexus of infrastructures and infrapolitical actions, as well as the effects of digitization on noncapitalist economies.

Platforms and Publics

Following Habermas's (2022) recent analysis of the platformization of the public sphere, communication scholars (e.g., see the contributions in Brunkhorst, Seeliger, and Sevignani 2024) have highlighted how the commercial nature of Facebook, Twitter, and other social media networks has compromised the processes of communication and will formation in liberal democracies. This is attributed to their attention-seeking business models, the degradation of discourse quality due to the absence of gatekeepers, as well as the media and journalism's adaptation to platform logic. As a result, they argue, the public sphere loses its capacity to stabilize society and to integrate, as it becomes increasingly difficult to transform deliberative processes into political action. The ambivalences and frictions these scholars describe as inherent to the digitalized public sphere, between democracy and the market, the private and the public, autonomy and heteronomy, and increasing but polarizing publics, however, play out completely differently in the Cuban context.

First of all, the profit-driven nature of social media platforms, characterized by advertising and data collection, has a diminished effect on the island due to the U.S. embargo, which prevents these corporate entities from commodifying communication or user data or deploying targeted ads. Conversely, digital platforms dedicated to sales and barter that have emerged on these sites now constitute significant alternatives to the state-controlled distribution system, arguably creating more, not less, citizen autonomy. Further, the paradoxical relationship between the public and private in the digital public sphere manifests distinctly in Cuba compared to Western liberal societies. Habermas and others have noted that the blending of private self-expression with public participation on social media, the facility with which information can be decontextualized and recontextualized, and the individual algorithmic curation determining the personal and public information users are exposed to have contributed to a deterioration of contextual integrity. This has undermined the deliberative ideal of the public sphere where collective will is formed through rational-critical debate over matters of mutual interest. In Cuba's authoritarian public sphere, critical debates have traditionally been confined to the private sphere (*la casa*), as the state exercises control over public spaces (*la calle*), where private citizens are prohibited from assembling as a public entity independent of state oversight. State policy, however, remains ambiguous as to whether the island's emerging networked public sphere counts as calle or casa, and it is this very ambiguity that has provided Cubans with increased latitude. Controlling online spaces poses a greater challenge for the Cuban state compared to authoritarian governments like China, Russia, or Turkey. Cuba lacks both the leverage to pressure platforms for access to their servers or user data and the technological capabilities to offer state-controlled domestic alternatives to Western social media platforms (the restrictions that Cuban users currently experience on these platforms therefore stem more from the provisions of the U.S. embargo). Consequently, during the July 11 protests, the Cuban government resorted to completely cutting off Internet access to hinder protestors' ability to organize. Since the government is unable to regulate content on social media, individuals have seized the opportunity to voice grievances and, in some cases, hold officials accountable. Moreover, digital connectivity has nurtured the "weak ties" among citizens, enabling exchanges over private matters such as barter exchanges or engaging in more "banal" forms of activism, thereby expanding the scope of civic engagement beyond former boundaries. The state has gradually come to tolerate these more participatory new political middle grounds, under the condition that the political system itself remains unchallenged and citizens avoid manifesting their grievances through street protests. In such cases, the state's highly efficient offline repression apparatus swiftly responds, as observed following the July 11 demonstrations.

Finally, contrary to Habermas's concerns about digital communication technology fragmenting the public sphere and leading to separate user communities and sub-publics, this tendency has had an empowering effect in Cuba. The emergence of micropublics through grassroots networks such as el paquete semanal or SNET, as well as numerous semiprivate chat groups on platforms like Facebook, WhatsApp, and Telegram, has fostered free spaces attracting diverse participants and engaging with distinct social issues. These platforms have allowed many Cuban citizens to experience significant autonomy. These varied yet interconnected arenas are not completely independent of state oversight, nor are they spaces of unfettered self-determination, and they may not directly lead to widespread political action, mass mobilization, or systemic change. Nevertheless, I assert that these arenas have the potential to serve as catalysts for long-term transformation, owing to the myriad negotiations and compromises that have shaped and defined them. Cubans' engagement with digital communication media is, therefore, not primarily motivated by political resistance, contrary to what many U.S. commentators in the wake of July 11 have suggested. Instead, it still upholds many of the state-promoted socialist values, conventions, and ideas. Although Cubans are undeniably frustrated by the ongoing material scarcities and economic woes, their political claims and concerns are diverse and significantly influenced by generational and socioeconomic differences, as I demonstrated throughout this book.

Infrastructure and Infrapolitics

This book has advanced an understanding of publics as collaboratively constructed, grounded in practical contexts, and shaped by sociotechnical infrastructures—a dynamic perspective that goes beyond the limitations of the static ontology inherent in the Habermasian model. Cubans' creation of vernacular infrastructure circuits, which have led to new communities and pathways for horizontal communication, exemplifies the reciprocal relationship between infrastructures and publics and highlights their intertwined roles in the sociotechnical organization of society. I have further argued that in authoritarian contexts such as Cuba, this infrastructuring of publics, the bottom-up creation of new communicative spaces, holds significant political implications. This remains true even if the discourse within these spaces does not directly engage with political subjects, and those involved in maintaining these spaces may not view their actions as inherently political. In environments where avenues for overt political engagement, such as widespread social media campaigns, are limited, activism tends to adopt more localized and everyday expressions, as Jeremy Morris, Andrei Semenov, and Regina Smyth (2023) have argued for the case of contemporary Russia. They note that activists in oppressive regimes often perceive their actions as apolitical, as the overwhelming and confrontational dynamics of power politics prompt citizens to disengage from formal political domains altogether. For many grassroots activists, politics is viewed as corrupt, resulting in the discreditation of the political sphere as a whole.

A recent surge in scholarship has examined the infrapolitical responses of citizens and activists to the global shift toward "networked authoritarianism." This includes not only Internet censorship in autocratic regimes but also increasingly repressive policing practices in democracies, where activities on digital platforms are monitored and citizens are targeted with facial recognition and other surveillance technologies. Ashley Lee's work (2018, 2022), for example, illustrates how young activists in Canada and the United States as well as in authoritarian Cambodia navigate the pressures of social, state, and corporate surveillance and other forms of digitally mediated control as they participate in social movements. She describes their under-the-radar tactics that involve the use of hidden or covert forms of expression, organizing within concealed subcultural spaces, networks, or channels, and adopting low-profile, disguised, anonymous, or pseudonymous identities online. Within the tightly controlled digital landscape of China, Yang Peidong, Lijun Tang, and Xuan Wang (2015) have highlighted how social media users employ coded forms of communication, including puns and neologisms, to mock or challenge government discourse. The Cuban example enriches these discussions of digital infrapolitics by showcasing a variety of practices that not merely creatively engage with the modalities of online communication but significantly expand the material space for civic discourse itself. Digitization in Cuba has enabled citizens to engage in practices of rerouting, hacking, and tinkering that all take place just beyond the surveillance of the state and often beyond, or sometimes atop, its ICT infrastructure. These activities represent a form of nonpoliticized activism, eschewing explicit political organizing and claims making while carefully avoiding actions that could draw scrutiny from state authorities. Digital infrapolitics in Cuba, deeply interwoven with material practices of infrastructuring, thus significantly expands our comprehension of the varied strategies of resistance accessible in the digital realm. Although the liberties generated by these strategies are often primarily economic, they are gradually and subtly altering the boundaries of permissible discourse.

Another Digital Economy

Last, the case of Cuba's digital revolution also contributes to our understanding of the impact of digitization on "other" economies that have been largely overlooked in interdisciplinary research on digital economic transformations. As Geoffrey Hobbis (2021) has argued, studies of "surveillance capitalism" (Zuboff 2019), "data colonialism" (Couldry and Mejias 2019), and "platform capitalism" (Srnicek 2017) presuppose that the adoption of digital technologies always results in an assimilation into the capitalist, free-market system. He therefore urges an engagement with the technological experiences of societies that maintain long-standing, contemporary alternative economies. In Cuba's current economic system, characterized by a complex mix of socialism and market reforms—including state control over key sectors, incremental openings for private businesses and foreign investment, and the transition from a long-standing dual currency system, which had created separate economic spheres, to dollarization—digitization has had multifaceted effects. Increased access to technology and corporate digital platforms has provided new economic possibilities for Cubans. Vernacular distribution infrastructures, such as el paquete, WhatsApp business accounts, and Telegram groups, all have facilitated platforms for the informal economy, resulting in greater freedom from governmental structures. However, Cubans' adoption of these platforms navigates a complex terrain, simultaneously upholding and subverting the values and conventions of the (post)revolutionary regime. This dynamic positions the moral economy of the users in continuous negotiation and friction both with the moral economy of the state and its socialist ideals as well as with the moral economy of the platforms, which are driven by capitalist logics that seek to commodify all social relations by collecting, algorithmically processing, and selling user data. These everyday moral negotiations among contrasting value systems surfaced, for instance, in the ongoing disputes between SNET's users and administrators regarding the network's dual identity as both a space for sharing and collaboration and a venue for unchecked profit generation or the uneasy coexistence of acts of solidarity with market-mediated exchanges within barter groups on WhatsApp and Telegram. The utilization of social media platforms by Cubans has further facilitated a renewed connection with the Cuban diaspora. This digital bolstering of transnational support networks has contributed to the ongoing transformation of the Cuban economy toward one increasingly driven by remittances.

A bottom-up moral economy shaped by ideals of solidarity is further enacted by actors such as the Copincha collective, who consciously utilize digital technology to establish alternative distribution networks for sharing books, tools, designs, and knowledge that enhance the resilience of Cubans' "other" economic practices and aim to disrupt the logics of digital capitalism. These examples illustrate that the capitalist domination of the digital realm is not as inevitable as many scholars suggest. Cuban users possess the agency for moral reasoning on digital platforms, enabling them to prioritize reciprocity and relationality as principles for their interactions. Hence, the Cuban example vividly highlights the social and relational importance of technology, effectively challenging the deterministic views on the impact of digital technologies prevalent in many studies of the digital economy and platforms.

The moral economies that influence the actions and values of people living in non- or not fully capitalist systems offer a wealth of examples of alternative economic arrangements. These systems are not motivated by capitalism's ceaseless quest for growth and value extraction, which is increasingly threatening the survival of our planet, and studying and understanding them is therefore more important than ever before. Despite the internal conflicts, external pressures, and material limitations they face, Cuba's grassroots technological infrastructures, in this regard, present distinct approaches to the Silicon Valley business model, whose favoring of maximizing profits for shareholders at the expense of users leads to ever-increasing "enshittification," as Cory Doctorow (2023) has famously phrased it. They illustrate how our access to and interaction with technology, along with the prevailing Western technoconsumer culture, could be reimagined. The decentralized distribution systems, embodied in people-embedded infrastructures like el paquete and SNET, allow for the dissemination of information and media content without dependence on centralized platforms controlled by governments or big tech corporations and thus conform better to what Doctorow terms the "end-to-end-principle," which advocates that networks should be designed to ensure that messages from willing speakers reach willing listeners' endpoints as swiftly and reliably as possible. In contrast to the surveillance-based revenue strategies of these companies that collect and monetize vast amounts of user data while seeking market dominance, these grassroots infrastructures present blueprints for less extractive, less environmentally destructive, more inclusive, and more horizontal means of connecting people.

While social media platforms lock in their users by prohibiting interoperability and increasingly make creators pay "ransom money" to have their content included in subscribers' feeds, el paquete, for instance, provides its users with equal and inclusive access to data and information, even bridging the digital divide for those who don't have regular or affordable Internet access. Instead of opaque machine-learning-based recommendation systems, targeted advertising, and algorithmic amplification of extreme content, this network primarily features locally created or curated content and can quickly adapt to user preferences, fostering transparency and accountability and prioritizing the needs and values of specific communities. In this regard, these Cuban digital practices mirror a growing global trend of users retreating from the hypervisibility and exploitation of mainstream platforms to smaller, more intimate digital spaces like Discord servers, Substack newsletters, Telegram channels, and invite-only forums—a trend encapsulated by the "Dark Forest" theory of the Internet, popularized by Yancey Strickler and others (Dark Forest Collective 2024). These spaces, often referred to as "Web 2.5," offer refuge from the algorithmic manipulation, data harvesting, and toxic dynamics of the "clearnet." In the Dark Forest, users prioritize meaningful, in-depth interactions over viral reach, fostering communities built on trust and shared values rather than surveillance and profit.

Furthermore, the community-owned infrastructure of SNET and other DIY computer networks empowers users to take control of their own technology, in contrast to residential Internet access provided by commercial Internet service providers (ISPs) in capitalist contexts that often constrain user autonomy. In the case of ISP customers, their ability to utilize the routers rented from their telecommunications provider to create their own networks is typically limited by imposed limitations such as port blocking and bandwidth throttling. This restrictive setup disempowers users and favors the ISP, as the user-provider relationship prioritizes consumption over user empowerment. Moreover, the asymmetrical nature of download speeds being faster than upload speeds perpetuates a dynamic where users predominantly consume content rather than actively contribute to the network. Furthermore, ISP customers generally have little stake in matters of data collection and surveillance, resulting in a lack of control over their personal information. This arrangement undermines user self-determination and privacy, as users are left with limited control over the collection and use of their data by ISPs.

The Cuban community-led initiatives I have discussed in this book thus exemplify bottom-up, autonomous, and sustainable local networks. By necessity, these initiatives utilize open hardware and software, leverage local physical and social infrastructures, and make use of reused materials or, in the case of Copincha's solar-powered microserver, even natural resources to enable communities to create and interconnect their own diverse networks, visions, and worlds. Therefore, they challenge the endless pursuit of innovation as an end in itself, which is based on a linear idea of progress and often leads to technological obsolescence and the prioritization of short-term economic gains above all else.

Such alternative approaches become even more crucial considering the relentless increase in the Internet's energy consumption, which now represents a progressively significant portion of global electricity usage, despite approximately 2.8 billion people still not having access to it (Statista 2023). For us future citizens of a broken world, it will therefore become increasingly important to create networks that are more resilient, more inclusive, more solarpunk, and thus more Cuban.

NOTES

Introduction. Cuba's Digital Awakening

1 For a detailed analysis of these reforms, see Mesa-Lago and Pérez-López (2013), Ritter and Henken (2015), Feinberg (2016).

2 However, most of these licensed professions require only a low level of qualification and therefore do not create competition for state-run enterprises.

3 Race and racial inequalities are complex subjects in Cuba. While people tend to discuss race more openly compared to in the United States or Europe, often describing or nicknaming others based on racial appearance, racism itself is seldom addressed. This limited discourse largely originates from the government's self-proclaimed achievement of eliminating racial injustices. While the Revolution has undoubtedly improved the life prospects of Afro-Cuban citizens, it remains evident that the highest echelons of the Cuban Communist Party are predominantly occupied by individuals of white descent.

4 Whereas Cuban consumers and tourists were subject to an official conversion rate of 24 CUP for 1 CUC, accounting and exchange operations in state-owned enterprises functioned at a rate of 1:1. This was unsustainable because it disguised accounting losses and surpluses and removed incentives to increase efficiency and productivity since economic results looked the same whether products were sold domestically for CUP or against hard currency on the global market, even if the monetary value to the Cuban government was significantly different (Yaffe 2021).

5 MLC is an abbreviation of *moneda libremente convertible* (freely convertible currency), a dollar-pegged virtual currency circulating only via bank cards. Hard currencies such as euros and U.S. and Canadian dollars are exchanged to MLC once they are transferred to an MLC account with a Cuban bank.

6 In August 2022, the Central Bank took measures to combat the black market and increase the flow of remittances through official banking channels. As part of this strategy, the official exchange rate of 1:24 and exchange unification were abandoned. Instead, a rate of 120 CUP per U.S. dollar was established, which closely mirrored the prevailing informal exchange rate at the time. This rate would be applicable to individual exchange transactions conducted through official exchange houses (Cadeca) and banks, including both cash and transfers from abroad. However, state-owned or foreign-invested companies as well as other types of organizations continued to operate at the rate of 24 CUP per dollar. This situation reintroduced the problematic policy of multiple exchange rates. According to economists like Pavel Vidal Alejandro and Omar Everleny Pérez Villaneuva (2022), this measure only expedited the depreciation of the Cuban peso, leading to further devaluation.

7 The date commemorates the attack on the army barracks on Santiago de Cuba on July 26, 1953, a first attempt at overthrowing the rule of dictator Fulgencio Batista. It also inspired the name of Fidel Castro's vanguard revolutionary organization (which later became a political party), the Movimiento 26 de Julio (26th of July Movement).

Chapter 1. Wi-Fi

1 Despite their self-declared humanitarian mission, Brothers to the Rescue operated in clear opposition to the Cuban government and also dropped dissident leaflets over Cuba. The Cuban government therefore accused the group of involvement in terrorist acts and even infiltrated them.

2 To this day, digital sovereignty is one of the central goals of the Cuban state and is expressed, for example, in the fact that it has developed numerous national alternatives to global Internet services.

3 Hoffmann (2014) argues indeed that offline repression was closely linked to the opening of the Internet from its very start. He sees the infamous report of the Politburo, which crushed these debates on a more autonomous civil society, as a prerequisite for the island's subsequent connection to the global Internet.

4 Another avenue for Cubans to access the Internet without surveillance was through the embassies of countries like the Czech Republic, Switzerland, and Spain. However, utilizing these connections could label individuals as oppositionists in the eyes of the Cuban state, as these countries actively supported dissidents and had policies aligned with promoting democratic values and human rights. Therefore, accessing the Internet through these channels implied a certain political stance and carried the risk of drawing scrutiny from Cuban authorities.

5 For an extended analysis of Cuban grassroots media practices as articulation work, see Dye 2019.

Chapter 2. Hard Drives

1 The term *maceta* translates to "flowerpot" and metaphorically referred to the act of uprooting and eliminating these illicit practices. Under Operación Maceta, the government took stricter measures to enforce regulations and control informal economic activities. It involved increased surveillance, raids, and arrests of individuals involved in activities such as smuggling, illegal trading, and other forms of underground economic endeavors. The goal was to maintain control over the economy and suppress unauthorized commercial activities that emerged during the economic crisis. The implementation of Operación Maceta reflected the Cuban government's efforts to protect the state-controlled economic system and maintain its socialist principles. It was part of a broader strategy to combat the adverse effects of the economic downturn and reinforce the central planning system in the face of growing informal markets and the emergence of a more market-oriented mindset among some Cubans (Mesa-Lago and Pérez-López 2005).

2 As of today, the majority of individuals involved in this human infrastructure, whether as matrices or as paqueteros, continue to operate legally under the *comprador-vendedor de discos* license (Köhn 2019). However, since 2017 the Cuban government has halted the issuance of this license. Many paqueteros whom I spoke with interpreted this decision as a concession to the United States during the period of political thawing in the Obama era. The widespread media piracy on the island was seen as an obstacle to the reestablishment of economic relations between Cuba and the United States.

3 Paqueteros told me that before the pandemic a one terabyte hard drive cost between seventy and eighty dollars on the black market. This price rose to more than one hundred when the border closings halted the influx of new consumer goods.

4 Both the Cuban state media and private casas matrices also share a common practice of disregarding international copyright laws. This disregard stems from a combination of ideological conviction and practical necessity. The Cuban state, driven by its ideological commitment to *bienes communes* (commons) and universal access to culture, places these principles above the capitalist property regime upheld by copyright laws. Additionally, the U.S. embargo on Cuba further complicates matters by making it impossible for Cuban state media to officially license material from American content providers, even if they desire to do so. Consequently, the Cuban government has long relied on piracy as a means to overcome these challenges. Examples of this piracy can be seen in the provision of reprinted foreign textbooks to Cuban universities, the use of pirated software to operate the country's official digital infrastructure, and the broadcast of foreign films

on state television, often with the burned-in inscription "for festival use only." Cuentapropistas such as paqueteros, or copy store owners, who sell copies of international media products within Cuba thus have every right to do so, despite knowingly violating international copyright laws. As long as they possess a national license, their business operations are deemed legitimate, and the state even benefits from their activities through the taxes they pay.

Chapter 3. Networks

1 Such conflicting concepts of the network as a site for sharing and collaboration versus the unbridled generation of profits in market exchanges also shape Western perspectives on the Internet. See, for example, Gabriella Coleman's (2012) work on hacker ethics and the free software movement.

2 In recent years, similar local practices of finding innovative and improvised solutions in response to a scarcity of resources, such as *gambiarra* in Brazil (Fonseca 2015) and *jugaad* in India (Rai 2019), have gained increasing anthropological interest.

3 However, since the integration of SNET into the government intranet, many former SNET administrators and users have succeed in the emerging Cuban private sector, where they are now applying the cultural capital and technical skills they acquired in creating and maintaining the network. Among them are, for example, the members of Cuba's most successful independent video game studio, ConWiro (with whom I developed the documentary video game *PakeTown*), and the webmasters of many small advertising or online promotion agencies. Some of them expressed the hope that in this way SNET could be the blueprint for the development of native Cuban platforms that in the future could offer local alternatives to the services of the big U.S. tech companies.

4 Even with better Internet access, Cuban gamers would still be excluded from using these global online game distribution platforms as the U.S. embargo prohibits U.S. companies from doing business in Cuba and also makes online payments from Cuba impossible.

5 Such linking was more important in SNET than on the Internet as SNET did not consistently use DNS (Domain Name System), which means that users couldn't access any page just by typing its domain name in their browser but had to know its IP address.

6 Some of these former Netlab members subsequently joined the Copincha hacker and makerspace to pursue their projects (see chapter 6).

7 This politics of integrating alternative areas of expression is embodied in Fidel Castro's famous dictum "within the Revolution; everything; against the Revolution, nothing" from his speech known as "Words to the Intellectuals" in June 1961 (Castro [1961] 1972, 276). For a case study of state attempts to incorporate Cuba's emergent hip-hop scene within the state's institutional structures, see Perry (2015, 172–97).

Chapter 4. Markets

1 Due to the unproductivity of the highly regulated agricultural sector, unfavorable climatic conditions, and lack of investment in machinery, Cuba must import 70 percent of the food the country consumes (Dominguez and Arencibia 2021).

2 Telegram, notably, operates with a revenue model distinct from that of WhatsApp and Facebook. It relies not on the sale of user data but instead on premium subscription services and in-app purchases. Despite a market valuation of $30 billion currently, it is not yet profitable and continues to be financed by the personal funds of its founder, Pavel Durov. The company has however laid out a clear strategy to monetize the platform in a way that aligns with its values and the commitments made to its users over the years.

3 The mass spread of the mobile Internet is also remarkable because the corporate state-owned monopoly ETECSA demands high prices. Access to the mobile Internet is paid for not by the hour but by data packages. At the outset, the cheapest mobile data package of 600 MB

per month cost 7 CUC, while the 4 GB package cost 30 CUC, equivalent to the average monthly wages for state employees before monetary union. In 2020, a 2.5 GB LTE package still cost 8 CUC.

4 One particular incident that vividly highlighted the inefficiency of the Cuban corporate state was when Nestor and I decided to abandon our shopping attempt after waiting for seven hours outside the Náutico supermarket in the upscale Playa district. A policeman responsible for managing the queue informed us that the digital payment system had repeatedly malfunctioned due to connection issues. As a result, only around a hundred customers had managed to make any purchases thus far that day.

5 Independent farming exists in Cuba, but within narrow legal limits. Producers are usually forced to deliver production quotas to the state and can then offer the rest on the market.

6 As a result of the new MIPyME laws enacted in August 2021 that legalized micro, small, and medium-sized enterprises, Havana saw the opening of small private grocery stores throughout 2023, after often lengthy approval processes. These private retail shops, mostly set up in garages or private homes, were the first of their kind since the Revolution and sell many of the basic supplies that were previously hard to find. The prices at these stores, which have to import their goods through state-owned agencies that take a 20 percent commission, are unaffordable for Cubans who earn state salaries. Consequently, online sales and barter platforms continue to exist.

7 Despite reforms to increase the private sector, in 2019 almost 70 percent of working Cubans still were employed by the state (ONEI 2020).

8 Before the monetary reorganization, a typical state salary ranged between 20 CUC (for unskilled work) and 160 CUC (for highly skilled workers in leading positions—e.g., in the medical sector) a month. After January 1, 2021, the state significantly raised these salaries (many job categories were increased by a factor or three or four), but at the same time slashed the subsidies on electricity, gas, or foodstuffs, which ultimately resulted in little to no improvement for much of the population.

9 For a more in-depth analysis of the aesthetics and use of photography in these online markets, see Köhn and Siré (2021).

10 These practices strongly echo the pioneering web trends of the late 1990s and early 2000s, which introduced transformative economic models like online gift economies, peer-to-peer (P2P) economies, and what Yochai Benkler (2006) described as "the wealth of networks." These developments seemed to facilitate the rise of previously marginal(ized) collaborative practices on a large scale. Such participatory sharing practices have, according to many critics, been undermined by platform capitalism. Instead of fostering a more equitable "sharing economy" as initially envisioned, platforms often mirror the structure and expectations of the venture capital investments supporting them, as Paul Langley and Andrew Leyshon (2017) argue. They frequently constrain true user participation by directing it toward activities that generate profit for the creators of the platforms.

11 Miguel Díaz-Canel Bermúdez, Twitter post, January 30, 2019, 1:11 PM, https://twitter .com/DiazCanelB.

Chapter 5. Publics

1 Despite potential delays of several months for the money to arrive, many hosts preferred cash payments as this made it more challenging for the state to properly tax them. Even when in 2017 Airbnb introduced AIS cards, issued by the Cuban financial institution Fincimex for receiving remittances from abroad as a new payment option, the preference for cash payments remained unchanged. However, the Cuban corporate state still benefited from the currency conversion from U.S. dollars to CUC, as recipients received only 0.87 CUC for each dollar due to the high taxes imposed by the government on transactions in U.S. dollars.

2 Same-sex marriage ultimately did become legal when the government held a public refer-
endum on a new "family law" code in September 2022. The measure that was heavily promoted by
the government was approved by 66.9 percent to 33.1 percent, having again met strong resistance
from the growing evangelical movement (Criado Pérez and Valdés Moreno 2023).

3 As part of the Cuban educational system, university graduates are required to work for
two to three years in a public institution as a form of repayment for their free education. Consequently, all young journalism graduates inevitably gain firsthand experience working with the
official state-controlled press.

4 The "Miami mafia" is a common refrain used by the Cuban government to portray exiles and
dissidents as a homogeneous group with malicious intentions. My friend used the term to distance
herself from the ideological and moral corruption she perceives on both sides of the Cuban divide.

5 José Jasán Nieves Cárdenas, personal communication.

6 Samantha Power, Twitter post, February 24, 2021, 11:14 PM, https://twitter.com/Samantha
JPower.

7 Julián Macías Tovar, Twitter post, July 12, 2021, 10:22 PM, https://twitter.com/Julian
MaciasT.

8 García purposefully chose to reference Martí's poem "Cultivo una rosa blanca" (I cultivate
a white rose), renowned for its profound message of forgiveness and the nurturing of virtues
like love, sincerity, and compassion, to symbolize national unity and avoid being perceived as
a dissident.

9 During the Mariel crisis, Fidel Castro temporarily allowed the opening of the Mariel port
as an exit point for Cubans wishing to emigrate. Over a period of seven months, approximately
125,000 Cubans took the opportunity to migrate to the United States. During the Balsero crisis,
over 35,000 Cubans embarked on makeshift rafts to cross the Straits of Florida in order to escape
the hardships of the Special Period.

10 After the Clinton administration's 1995 revision of the Cuban Adjustment Act (U.S. Congress 1966, amended 1995), the special privileges granted under the act were limited to Cubans who
arrived in the United States by land, commonly known as having "dry feet." Those intercepted at
sea, referred to as having "wet feet," were subject to repatriation. Obama then ended Clinton's "wet
foot, dry foot" policy in 2017 during his last week in office. This formally subjected Cuban migrants
to the same treatment as migrants from other countries. Yet in practice Cuban exceptionalism
persisted because Obama did not seek to persuade the Republican-controlled Congress at the
time to repeal the Cuban Adjustment Act entirely, allowing the continuation of certain benefits.
In February 2023, the Biden administration implemented further changes to Cuban privileges in
response to the perceived migration crisis at the U.S.-Mexican border. As a result, Cubans were
now at risk of expulsion upon unauthorized entry, with a five-year ban on reentry, and they were
no longer eligible to claim asylum if they had passed through a third country. However, in an
effort to provide alternative legal pathways, the administration expanded the bilateral immigration accord that had been previously disregarded by the Trump administration. This included a
promise to issue 20,000 visas annually for Cubans. Additionally, the administration announced
plans to reactivate consular services in Havana and launched a new humanitarian parole program.
Under this program, up to 30,000 Cuban, Venezuelan, Haitian, and Nicaraguan nationals who had
a sponsor in the United States would be admitted each month. In May 2023, the Department of
Homeland Security reported that 380,000 parole applications for Cubans were pending review.
Indeed, during my trip to the island in December 2023, it felt like virtually all my acquaintances
who hadn't yet left were continuously monitoring the status of their parole application on the U.S.
Customs and Border Protection's CBP One mobile app. On January 24, 2025, only days after his
second inauguration, President Trump canceled the humanitarian parole programs for individuals

from Cuba, Haiti, Nicaragua, and Venezuela through an executive order titled "Securing Our Borders." In June 2025, the Department of Homeland Security began issuing termination notices to participants and revoking associated work permits. How his policies will further impact Cubans' migration privileges remains to be seen.

Chapter 6. Repair

1 The Oficina del Historiador was created before the Revolution, in 1938, for the protection of the architectural heritage of Habana Vieja. The institution oversees the restoration of the city's historic center, which was declared a UNESCO World Heritage Site in 1982. It has its own budget, which is mainly financed by tourism revenue, and can therefore act with a certain independence from the government. After the 27N protests (described in chapter 5), many of my artist friends refused to participate in exhibitions or activities organized by institutions belonging to the Ministry of Culture but would still get involved with projects run by the Oficina del Historiador as this institution has more political leeway.

2 Since the acidic soils in many parts of the tropics do not permit long-term cultivation, the fields remained productive for a few years and were then abandoned in favor of a new plot. The great abundance of forests in the pre-Columbian period allowed nature to naturally reclaim the land in use. Hence, this type of slash-and-burn cultivation, as a cyclical farming method that utilizes the regenerative power of nature, was perfectly suited to the Caribbean soil and climate.

3 This endeavor yielded tangible results, for example, when Copincha started producing face shields for the medical sector on behalf of the state during the pandemic. Their efforts proved successful and had a notable impact, prompting the government to eventually ease restrictions on the importation of 3D printers.

BIBLIOGRAPHY

Aja Díaz, Antonio, and María Ofelia Rodríguez Soriano. 2022. "Apuntes Para La Evaluación de La Migración Internacional de Cuba." *Revista Novedades En Población* 18 (36): 1–32.

Allen, Jafari. 2012. "One Way or Another: Erotic Subjectivity in Cuba." *American Ethnologist* 39 (2): 325–38. https://doi.org/10.1111/j.1548-1425.2012.01367.x.

Alonso, Martín Oller, and José Raúl Concepción Llanes. 2020. "Cultural Studies in Latin America: 'Packaged Cuba.'" In *Off and Online Journalism and Corruption—International Comparative Analysis*, edited by Basyouni Ibrahim Hamada and Saodah Wok. London: IntechOpen. https://doi.org/10.5772/intechopen.85920.

Alzugaray, Carlos. 2021. "The 11J Demonstrations in Cuba: A Provisional Assessment." In *The Road Ahead—Cuba after the July 11 Protests*, edited by William M. LeoGrande, John M. Kirk, and Philip Brenner. Washington, DC: American University. https://www.american.edu/centers/latin-american-latino-studies/cuba-after-the-july-11-protests-alzugaray.cfm.

Anand, Nikhil. 2017. *Hydraulic City: Water and the Infrastructures of Citizenship in Mumbai.* Durham, NC: Duke University Press.

Anand, Nikhil, Akhil Gupta, and Hannah Appel, eds. 2018. *The Promise of Infrastructure.* Durham, NC: Duke University Press.

Andaya, Elise. 2009. "The Gift of Health." *Medical Anthropology Quarterly* 23 (4): 357–74. https://doi.org/10.1111/j.1548-1387.2009.01068.x.

Angel, Sergio, Claudia González, María Matienzo Puerto, Ana Paula López, Nelson Álvarez, Louis Thiemann, and Alejandra Suárez. 2020. *Formas de sobrevivencia en cuba: "resistencias cotidianas" en la habana, matanzas y sagua la grande.* Bogotá: Universidad Sergio Arboleda. https://repository.usergioarboleda.edu.co/handle/11232/1692.

Aouragh, Miriyam. 2015. "Revolutions, the Internet and Orientalist Reminiscence." In *Revolutionary Egypt*, edited by Abou-El-Fadl Reem, 257–278. Abingdon: Routledge.

Assange, Julian. 2008. "Cuba to Work around US Embargo via Undersea Cable to Venezuela." *Wikileaks*, July 16, 2008. http://wikileaks.org/wiki/Cuba_to_work_around_US_embargo_via_undersea_cable_to_Venezuela.

Barbrook, Richard, and Andy Cameron. 1996. "The Californian Ideology." *Science as Culture* 6 (1): 44–72. https://doi.org/10.1080/09505439609526455.

Bastian, Hope. 2018. *Everyday Adjustments in Havana: Economic Reforms, Mobility, and Emerging Inequalities.* Lanham, MD: Lexington Books.

Bastian, Hope, and Maya J. Berry. 2022. "Moral Panics, Viral Subjects: Black Women's Bodies on the Line during Cuba's 2020 Pandemic Lockdowns." *Journal of Latin American and Caribbean Anthropology* 27 (1–2): 16–36. https://doi.org/10.1111/jlca.12587.

Benkler, Yochai. 2006. *The Wealth of Networks: How Social Production Transforms Markets and Freedom.* New Haven, CT: Yale University Press.

Bennett, W. Lance, and Alexandra Segerberg. 2013. *The Logic of Connective Action: Digital Media and the Personalization of Contentious Politics.* Cambridge: Cambridge University Press. https://doi.org/10.1017/CBO9781139198752.

Berg, Mette Louise. 2015. "'La Lenin Is My Passport': Schooling, Mobility and Belonging in Socialist Cuba and Its Diaspora." *Identities* 22 (3): 303–17. https://doi.org/10.1080/1070289X.2014.939189.

Boas, Taylor. 2000. "The Dictator's Dilemma? The Internet and US Policy toward Cuba." *Washington Quarterly* 23 (3): 57–67. https://doi.org/10.1162/016366000561178.

Bonini, Tiziano, and Emiliano Trere. 2024. *Algorithms of Resistance: The Everyday Fight against Platform Power*. Cambridge, MA: MIT Press.

Boyer, Dominic, and Alexei Yurchak. 2010. "AMERICAN STIOB: Or, What Late-Socialist Aesthetics of Parody Reveal about Contemporary Political Culture in the West." *Cultural Anthropology* 25 (2): 179–221. https://doi.org/10.1111/j.1548-1360.2010.01056.x.

Brito Chávez, Yecenia. 2014. "El Consumo Cultural Del Paquete En Jóvenes Universitarios." Thesis, Universidad Central "Marta Abreu" de Las Villas, Santa Clara.

Brunkhorst, Hauke, Martin Seeliger, and Sebastian Sevignani, eds. 2024. "Structural Transformation of the Public Sphere." Special issue, *Philosophy & Social Criticism* 50 (1).

Burrell, Jenna. 2011. "User Agency in the Middle Range: Rumors and the Reinvention of the Internet in Accra, Ghana." *Science, Technology, & Human Values* 36 (2): 139–59. https://doi.org/10.1177/0162243910366148.

Bustamante, Michael J. 2021. "11J, 'Patria y Vida,' and the (Not So) New Cuban Culture Wars." In *The Road Ahead—Cuba after the July 11 Protests*, edited by William M. LeoGrande, John M. Kirk, and Philip Brenner. Washington, DC: American University. https://www.american.edu/centers/latin-american-latino-studies/cuba-after-the-july-11-protests-bustamante.cfm.

Butler, Desmond, Jack Gillum, and Alberto Arce. 2014. "US Secretly Built 'Cuban Twitter' to Stir Unrest." *AP News*, April 3, 2014. https://apnews.com/article/technology-cuba-united-states-government-904a9a6a1bcd46cebfc14bea2ee30fdf.

Campbell Romero, Bryan. 2021. "Have You Heard, Comrade? The Socialist Revolution Is Racist Too." North American Congress on Latin America, August 9, 2021. https://nacla.org/cuban-socialist-revolution-racist-too.

Cárdenas, Harold. 2021. "The Generation Gap." In *The Road Ahead—Cuba after the July 11 Protests*, edited by William M. LeoGrande, John M. Kirk, and Philip Brenner. Washington, DC: American University. https://www.american.edu/centers/latin-american-latino-studies/cuba-after-the-july-11-protests-cárdenas.cfm.

Castells, Manuel. 2015. *Networks of Outrage and Hope: Social Movements in the Internet Age*. Hoboken, NJ: John Wiley.

Castro, Fidel. [1961] 1972. "Words to the Intellectuals." In *Radical Perspectives in the Arts*, edited by Lee Baxandall, 267–300. Harmondsworth: Penguin.

Cearns, Jennifer. 2019. "The 'Mula Ring': Material Networks of Circulation through the Cuban World." *Journal of Latin American and Caribbean Anthropology* 24 (4): 864–90. https://doi.org/10.1111/jlca.12439.

Centeno, Ramón I. 2017. "The Cuban Regime after a Decade of Raúl Castro in Power." *Mexican Law Review* 9 (2): 99–126. https://doi.org/10.22201/iij.24485306e.2017.18.10777.

Chalfin, Brenda. 2017. "'Wastelandia': Infrastructure and the Commonwealth of Waste in Urban Ghana." *Ethnos* 82 (4): 648–71.

———. 2023. *Waste Works: Vital Politics in Urban Ghana*. Durham, NC: Duke University Press.

Chase, Michelle. 2015. *Revolution within the Revolution: Women and Gender Politics in Cuba, 1952–1962*. Chapel Hill: University of North Carolina Press.

Coleman, E. Gabriella. 2012. *Coding Freedom*. Princeton, NJ: Princeton University Press.

Collier, Stephen J., James Christopher Mizes, and Antina von Schnitzler. 2016. "Public Infrastructures / Infrastructural Publics." *Limn*, November 9, 2016. https://limn.it/articles/preface-public-infrastructures-infrastructural-publics/.

Collins, Samuel Gerald, Matthew Durington, and Harjant Gill. 2017. "Multimodality: An Invitation." *American Anthropologist* 119 (1): 142–46. https://doi.org/10.1111/aman.12826.

Colomé, Carla Gloria. 2021. "11 de julio en San Antonio de los Baños: Lo que se ve/lo que no se ve." *El Estornudo*, July 22, 2021. https://revistaelestornudo.com/san-antonio-de-los-banos -protestas-11-julio-cuba/.

Concepción Llanes, José Raúl. 2015. "La Cultura Empaquetada. Análisis de Las Prácticas de Consumo Cultural Informal En Jóvenes Habaneros." Thesis, Universidad de La Habana.

Concepción Llanes, José Raúl, and Martín Oller Alonso. 2019. "'Repackaging' Cuban Cultural Consumption." *Lumina* 13 (2): 40–54. https://doi.org/10.34019/1981-4070.2019.v13.27743.

Couldry, Nick, and Ulises A. Mejias. 2019. *The Costs of Connection: How Data Is Colonizing Human Life and Appropriating It for Capitalism.* Stanford, CA: Stanford University Press.

Criado Pérez, Kirenia, and Dachelys Valdés Moreno. 2023. "Evangelical Christianity, the State, and New Political Alliances." In *Contemporary Cuba: The Post-Castro Era,* edited by Hope Bastian, Philip Brenner, John M. Kirk, and William M. LeoGrande, 307–12. Lanham, MD: Rowman & Littlefield.

Cubero, Carlo A. 2021. "What Does Anonymity Mean in Anthropological Filmmaking?" In *Rethinking Pseudonyms in Ethnography*, edited by Carole McGranahan and Erica Weiss. *American Ethnologist*, December 13, 2021. https://americanethnologist.org/features/collections/rethinking -pseudonyms-in-ethnography/what-does-anonymity-mean-in-anthropological-filmmaking.

Dalakoglou, Dimitris. 2012. "'The Road from Capitalism to Capitalism': Infrastructures of (Post) Socialism in Albania." *Mobilities* 7 (4): 571–86.

———. 2016. "Infrastructural Gap: Commons, State and Anthropology." *City* 20 (6): 822–31. https://doi.org/10.1080/13604813.2016.1241524.

Dark Forest Collective. 2024. *The Dark Forest Anthology of the Internet.* New York: Metalabel.

Dattatreyan, Ethiraj Gabriel, and Isaac Marrero-Guillamón. 2019. "Introduction: Multimodal Anthropology and the Politics of Invention." *American Anthropologist* 121 (1): 220–28. https:// doi.org/10.1111/aman.13183.

De Boeck, Filip. 2012. "Infrastructure: Commentary from Filip De Boeck. Contributions from Urban Africa towards an Anthropology of Infrastructure." *Cultural Anthropology*, November 26, 2012. https://journal.culanth.org/index.php/ca/infrastructure-filip-de-boeck.

De Boeck, Filip, and Marie-Françoise Plissart. 2004. *Kinshasa: Tales of the Invisible City.* Leuven: Leuven University Press.

De Filippi, Primavera, and Félix Tréguer. 2015. "Expanding the Internet Commons: The Subversive Potential of Wireless Community Networks." *Journal of Peer to Peer Production* 6: 1–40.

De la Fuente, Alejandro. 2001. *A Nation for All: Race, Inequality, and Politics in Twentieth-Century Cuba.* Chapel Hill: University of North Carolina Press.

Didion, Joan. 1987. *Miami.* New York: Simon & Schuster.

Dijck, José van, Thomas Poell, and Martijn de Waal. 2018. *The Platform Society.* Oxford: Oxford University Press. https://doi.org/10.1093/oso/9780190889760.001.0001.

Doctorow, Cory. 2023. "The 'Enshittification' of TikTok. Or How, Exactly, Platforms Die." *Wired*, January 23, 2023. https://www.wired.com/story/tiktok-platforms-cory-doctorow/.

Dominguez, Jessica, and Jesús Arencibia. 2021. "El Drama de La Comida En Cuba." *El Toque*, June 25, 2021. https://alimentoscuba.eltoque.com/.

Donner, Jonathan. 2015. *After Access: Inclusion, Development, and a More Mobile Internet.* Cambridge, Mass.: MIT Press. https://doi.org/10.7551/mitpress/9740.001.0001.

Dukalskis, Alexander. 2017. *The Authoritarian Public Sphere: Legitimation and Autocratic Power in North Korea, Burma, and China.* Abingdon: Routledge.

Duong, Paloma. 2013. "Bloggers Unplugged: Amateur Citizens, Cultural Discourse, and Public Sphere in Cuba." *Journal of Latin American Cultural Studies* 22 (4): 375–97. https://doi.org /10.1080/13569325.2013.840277.

Dye, Michaelanne. 2019. *Vamos a Resolver: Collaboratively Configuring the Internet in Havana.* Ph.D. Dissertation, Georgia Institute of Technology.

Dye, Michaelanne, David Nemer, Josiah Mangiameli, Amy S. Bruckman, and Neha Kumar. 2018. "El Paquete Semanal: The Week's Internet in Havana." Proceedings of the SIGCHI Conference on Human Factors in Computing Systems (CHI '18): 1–12.

Dye, Michaelanne, David Nemer, Neha Kumar, and Amy S. Bruckman. 2019. "If It Rains, Ask Grandma to Disconnect the Nano: Maintenance & Care in Havana's StreetNet." *Proceedings of the ACM on Human-Computer Interaction* 3 (CSCW): 187:1–187:27. https://doi.org/10 .1145/3359289.

Eaton, Tracey. 2021. "U.S. Government Democracy Projects in Cuba: Following the Money." In *The Road Ahead—Cuba after the July 11 Protests,* edited by William M. LeoGrande, John M. Kirk, and Philip Brenner. Washington, DC: American University. https://www.american.edu /centers/latin-american-latino-studies/cuba-after-the-july-11-protests-eaton.cfm.

Echemendía Pérez, Isabel M. 2015. "'Copi@ y Comp@rte Una Vez a La Semana'—Acercamiento a Los Principales Rasgos Que Caracterizan El Consumo Audiovisual Informal Del Paquete Semanal En Dos Grupos de Jóvenes de La Capital de Mayabeque." Thesis, Universidad de La Habana.

Eckstein, Susan. 1994. *Back from the Future: Cuba Under Castro.* Princeton, NJ: Princeton University Press.

———. 2022. *Cuban Privilege: The Making of Immigrant Inequality in America.* Cambridge: Cambridge University Press.

El Toque. 2022. "La Matrioshka de los negocios cubanos." May 6, 2022. https://eltoque.com /matrioshka-militar-cubana.

Elyachar, Julia. 2010. "Phatic Labor, Infrastructure, and the Question of Empowerment in Cairo." *American Ethnologist* 37 (3): 452–64. https://doi.org/10.1111/j.1548-1425.2010.01265.x.

EmmaStyle, dir. 2022. "Me Fui De CUBA🇨🇺Así Fue La Travesía De NICARAGUA a EE.UU 🇺🇸!!" https://www.youtube.com/watch?v=x4t38uNmRSo.

Estalella, Adolfo, and Tomás Sánchez Criado. 2018. *Experimental Collaborations: Ethnography through Fieldwork Devices.* New York: Berghahn Books.

Farber, Samuel. 2011. *Cuba since the Revolution of 1959: A Critical Assessment.* Chicago: Haymarket Books.

Favero, Paolo S. H., and Eva Theunissen. 2018. "With the Smartphone as Field Assistant: Designing, Making, and Testing EthnoAlly, a Multimodal Tool for Conducting Serendipitous Ethnography in a Multisensory World." *American Anthropologist* 120 (1): 163–67.

Feinberg, Richard. 2016. *Open for Business: Building the New Cuban Economy.* Washington: Brookings Institution Press.

Fernandes, Sujatha. 2006. *Cuba Represent! Cuban Arts, State Power, and the Making of New Revolutionary Cultures.* Durham, NC: Duke University Press.

Florida International University, Steven J. Green School of International & Public Affairs. 2022. "Cuba Poll—How Cuban Americans in South Florida View U.S. Policies Toward Cuba, Critical National Issues and the Upcoming Elections." https://issuu.com/fiupublications/docs /sipa_cuba_poll_report_2022_2882279691_final_noblee.

Fonseca, Felipe Schmidt. 2015. "Gambiarra: Repair Culture." *Tvergastein* 6 (1): 54–63.

Font, Mauricio A., and Alfonso W. Quiroz. 2005. *Cuban Counterpoints: The Legacy of Fernando Ortiz.* Lanham, MD: Lexington Books.

Fraser, Nancy. 1990. "Rethinking the Public Sphere: A Contribution to the Critique of Actually Existing Democracy." *Social Text* 25/26: 56–80. https://doi.org/10.2307/466240.

Fredericks, Rosalind. 2018. *Garbage Citizenship: Vital Infrastructures of Labor in Dakar, Senegal.* Durham, NC: Duke University Press.

García Santamaria, Sara. 2021. "Independent Journalism in Cuba between Fantasy and the On-tological Rupture." In *Cuba's Digital Revolution: Citizen Innovation and State Policy*, edited by Ted A. Henken and Sara García Santamaria, 180–99. Miami: University Press of Florida. https://doi.org/10.2307/j.ctvlmvw94t.

Garth, Hanna. 2020. *Food in Cuba: The Pursuit of a Decent Meal*. Stanford, CA: Stanford University Press.

Geoffray, Marie Laure. 2015. "Transnational Dynamics of Contention in Contemporary Cuba." *Journal of Latin American Studies* 47 (2): 223–49.

———. 2021. "Digital Critique in Cuba." In *Cuba's Digital Revolution: Citizen Innovation and State Policy*, edited by Ted A. Henken and Sara García Santamaria, 136–56. Miami: University Press of Florida. https://doi.org/10.2307/j.ctvlmvw94t.

Ghonim, Wael. 2012. *Revolution 2.0: The Power of the People Is Greater Than the People in Power: A Memoir*. Boston: Houghton Mifflin Harcourt.

Gibson, William. 1981. "Johnny Mnemonic." *Omni* 32: 56–63, 98–99.

———. 1982. "Burning Chrome." *Omni* 46: 72–77, 102–7.

Ginsburg, Faye. 2018. "Decolonizing Documentary On-Screen and Off: Sensory Ethnography and the Aesthetics of Accountability." *Film Quarterly* 72 (1): 39–49. https://doi.org/10.1525/FQ.2018.72.1.39.

Gold, Marina. 2015. *People and State in Socialist Cuba*. New York: Palgrave Macmillan. https://doi.org/10.1057/9781137539830.

González, Claudia. 2020. "Memes, sátiras y tropos en Cuba: humor digital como infrapolítica en la postrevolución." *Revista Foro Cubano* 1 (1): 3–22. https://doi.org/10.22518/jour.rfc/2021.1a01.

Götz, Norbert. 2015. "'Moral Economy': Its Conceptual History and Analytical Prospects." *Journal of Global Ethics* 11 (2): 147–62. https://doi.org/10.1080/17449626.2015.1054556.

Grenier, Guillermo J. 2023. "The March of Thousands: Some Considerations on the Incorporation of New Cuban Americans into the South Florida Political Culture." *OnCubaNews English* (blog), May 10, 2023. https://oncubanews.com/en/opinion/columns/it-is-not-easy/the-march-of-thousands-some-considerations-on-the-incorporation-of-new-cuban-americans-into-the-south-florida-political-culture/.

Guevara, Che. 1971. "Man and Socialism in Cuba." In *Man and Socialism in Cuba: The Great Debate*, edited by Bertram Silverman, 335–410. New York: Atheneum.

Guevara, Yurisander. 2015. "WiFi en el Ambiente. Juventud Rebelde informa en exclusiva sobre la apertura de 35 zonas de navegación por Internet con tecnología wifi en todo el país." *Juventud Rebelde*, June 15, 2015. https://www.juventudrebelde.cu/suplementos/informatica/2015-06-17/wifi-en-el-ambiente?page=1.

Habermas, Jürgen. 1991. *The Structural Transformation of the Public Sphere: An Inquiry into a Category of Bourgeois Society*. Cambridge, Mass.: MIT Press.

———. 2022. *Ein neuer Strukturwandel der Öffentlichkeit und die deliberative Politik*. Frankfurt a.M.: Suhrkamp Verlag.

Hansing, Katrin. 2017. "Race and Inequality in the New Cuba: Reasons, Dynamics, and Manifestations." *Social Research* 84 (2): 331–49. https://doi.org/10.1353/sor.2017.0022.

Hansing, Katrin, and Bert Hoffmann. 2020. "When Racial Inequalities Return: Assessing the Restratification of Cuban Society 60 Years after Revolution." *Latin American Politics and Society* 62 (2): 29–52.

Härkönen, Heidi. 2016. *Kinship, Love, and Life Cycle in Contemporary Havana, Cuba*. New York: Palgrave Macmillan. https://doi.org/10.1057/978-1-137-58076-4.

Harvey, Penelope, Casper Bruun Jensen, and Atsuro Morita, eds. 2016. *Infrastructures and Social Complexity: A Companion*. New York: Routledge.

Henken, Ted A. 2011. "Una Cartografía de La Blogósfera Cubana: Entre «oficialistas» y «merce-narios»." *Nueva Sociedad* 235: 90–109.

———. 2017. "Cuba's Digital Millennials: Independent Digital Media and Civil Society on the Island of the Disconnected." *Social Research* 84 (2): 429–56.

———. 2021a. "The Opium of the Paquete." *Cuban Studies* 50: 111–38.

———. 2021b. "From Generación Y to 14ymedio. Beyond the Blog on Cuba's Digital Frontier." In *Cuba's Digital Revolution: Citizen Innovation and State Policy*, edited by Ted A. Henken and Sara García Santamaria, 157–79. Miami: University Press of Florida. https://doi.org/10 .2307/j.ctvlmvw94t.

———. 2021c. "Del Movimiento San Isidro a 'Patria y Vida': ¿Quién Controlará La Revolución Digital Cubana?" *Revista Foro Cubano* 2 (2): 73–92.

Henken, Ted A., and Sjamme van de Voort. 2013. "From Nada to Nauta: Internet Access and Cyber-Activism in a Changing Cuba." *Cuba in Transition* 23: 341–50.

Hernández, Rafael. 2021. "Anatomía Del 27N Cubano y Su Circunstancia." *Nueva Sociedad*, Janu-ary 11, 2021. https://nuso.org/articulo/anatomia-del-27n-cubano-y-su-circuntancia/.

Hobbis, Geoffrey. 2021. "Digitizing Other Economies: A Critical Review." *Geoforum* 126: 306–9. https://doi.org/10.1016/j.geoforum.2021.08.006.

Hobbis, Geoffrey, and Stephanie Ketterer Hobbis. 2022. "Beyond Platform Capitalism: Critical Perspectives on Facebook Markets from Melanesia." *Media, Culture & Society* 44 (1): 121–40. https://doi.org/10.1177/01634437211022714.

Hoffmann, Bert. 2004. *The Politics of the Internet in Third World Development: Challenges in Con-trasting Regimes with Case Studies of Costa Rica and Cuba*. Abingdon: Routledge.

———. 2011. "Civil Society 2.0? How the Internet Changes State-Society Relations in Author-itarian Regimes: The Case of Cuba." GIGA Working Paper 156. https://www.econstor.eu /bitstream/10419/47847/1/655581456.pdf.

———. 2014. "Claiming Citizenship: Web-Based Voice and Digital Media in Socialist Cuba." In *Digital Technologies for Democratic Governance in Latin America*, edited by Anita Breuer and Yanina Welp, 200–216. Abingdon: Routledge.

———. 2016. "Bureaucratic Socialism in Reform Mode: The Changing Politics of Cuba's Post-Fidel Era." *Third World Quarterly* 37 (9): 1730–44. https://doi.org/10.1080/01436597.2016.1166050.

———, ed. 2021. *Social Policies and Institutional Reform in Post-COVID Cuba*. Leverkusen-Opladen: Verlag Barbara Budrich.

Holbraad, Martin. 2014. "Revolución o Muerte: Self-Sacrifice and the Ontology of the Cuban Revolution." *Ethnos* 79 (3): 365–87.

———. 2017. "Money and the Morality of Commensuration: Currencies of Poverty in Post-Soviet Cuba." *Social Analysis* 61 (4): 81–97. https://doi.org/10.3167/sa.2017.610406.

———. 2018. "'I Have Been Formed in This Revolution': Revolution as Infrastructure, and the People It Creates in Cuba." *Journal of Latin American and Caribbean Anthropology* 23 (3): 478–95. https://doi.org/10.1111/jlca.12344.

Holmes, Douglas R., and George E. Marcus. 2008. "Collaboration Today and the Re-imagination of the Classic Scene of Fieldwork Encounter." *Collaborative Anthropologies* 1 (1): 81–101. https:// doi.org/10.1353/cla.0.0003.

Horton, Chelsea. 2020. "Cuban Food Security in a Time of COVID-19." *Anthropology News* (blog), September 25, 2020. https://www.anthropology-news.org/articles/cuban-food-security-in -a-time-of-covid-19/.

Howe, Cymene, Jessica Lockrem, Hannah Appel, Edward Hackett, Dominic Boyer, Randal Hall, Matthew Schneider-Mayerson, Albert Pope, Akhil Gupta, Elizabeth Rodwell, Andrea Balles-tero, Trevor Durbin, Farès el-Dahdah, Elizabeth Long, and Cyrus Mody. 2016. "Paradoxical In-frastructures: Ruins, Retrofit, and Risk." *Science, Technology, & Human Values* 41 (3): 547–65.

Humphrey, Caroline. 2002. *The Unmaking of Soviet Life: Everyday Economies after Socialism*. Ithaca, NY: Cornell University Press.

———. 2005. "Ideology in Infrastructure: Architecture and Soviet Imagination." *Journal of the Royal Anthropological Institute* 11 (1): 39–58.

Humphreys, Laura-Zoe. 2019. *Fidel between the Lines: Paranoia and Ambivalence in Late Socialist Cuban Cinema*. Durham, NC: Duke University Press.

———. 2021. "Loving Idols: K-Pop and the Limits of Neoliberal Solidarity in Cuba." *International Journal of Cultural Studies* 24 (6): 1009–26.

Jackson, Steven J. 2014. "Rethinking Repair." In *Media Technologies: Essays on Communication, Materiality, and Society*, edited by Tarleton Gillespie, Pablo Boczkowski, and Kirsten Foot, 221–39. Cambridge, Mass.: MIT Press.

Jiménez Enoa, Abraham. 2018. "Los repatriados de Cuba: por qué miles de emigrantes están volviendo a la isla." *BBC News Mundo*, February 20, 2018. https://www.bbc.com/mundo/noticias-america-latina-43038889.

Jungnickel, Katrina. 2014. *DiY WiFi: Re-Imagining Connectivity*. Basingstoke: Palgrave Macmillan.

Kapcia, Antoni. 2008. *Cuba in Revolution: A History since the Fifties*. London: Reaktion Books.

Kelly, Sanja, Mai Truong, Madeline Earp, Laura Reed, Adrian Shahbaz, and Ashley Greco-Stoner. 2013. "Freedom on the Net 2013. A Global Assessment of Internet and Digital Media." Freedom House, October 3, 2013. https://freedomhouse.org/sites/default/files/2020-02/FOTN%202013_Full%20Report.pdf.

Khrustaleva, Olga. 2021. "Information and Communication Technology, State Power, and Civil Society: Cuban Internet Development in the Context of the Normalization of Relations with the United States." In *Cuba's Digital Revolution: Citizen Innovation and State Policy*, edited by Ted A. Henken and Sara García Santamaria, 73–94. Miami: University Press of Florida.

Kirk, John M. 2015. "Surfing Revolico.com: Cuba's Answer to Craig's List." In *A Contemporary Cuba Reader: The Revolution under Raúl Castro,* edited by Philip Brenner, Marguerite Rose Jiménez, John M. Kirk, and William M. LeoGrande, 443–46. Lanham, MD: Rowman & Littlefield.

Knorr, Alexander. 2009. "Maxmod: An Ethnography of Cyberculture." Habilitation thesis, Ludwig Maximilian Universität München.

Köhn, Steffen. 2019. "Unpacking El Paquete." *Digital Culture & Society* 5 (1): 105–24.

———. 2025. "The Screen Walk Method. Exploring the Social in/of Digital Environments." In *Empirical Art: Filmmaking and Theory-Making in the Social Sciences*, edited by Andy Lawrence and Martha-Cecilia Dietrich: 33–51. Manchester: Manchester University Press.

Köhn, Steffen, and Nestor Siré. 2021. "Messaging Groups and the Digital Black-Market in Cuba." *Unthinking Photography* (blog), December 11, 2021. https://unthinking.photography/articles/messaging-groups-and-the-digital-black-market-in-cuba.

———. 2022a. "Fragile Connections: Community Computer Networks, Human Infrastructures, and the Consequences of Their Breakdown in Havana." *American Anthropologist* 124 (2): 383–98.

———. 2022b. "Swap It on WhatsApp: The Moral Economy of Informal Online Exchange Networks in Contemporary Cuba." *Journal of Latin American and Caribbean Anthropology* 27 (1–2): 80–100.

———. 2022c. "Screen Walks: Conducting 'Research in Motion' in Digital Environments." *American Anthropologist*, April 16, 2022. https://www.americananthropologist.org/online-content/screen-walks.

———. 2023. "Sneakernets and Copy Houses: A Video Essay on Cuban Offline Culture." *Visual and New Media Review—Fieldsights*, September 7, 2023. https://culanth.org/fieldsights/sneakernets-and-copy-houses-a-video-essay-on-cuban-offline-culture.

Korn, Matthias, Wolfgang Reißmann, Tobias Röhl, and David Sittler. 2019. "Infrastructuring Publics: A Research Perspective." In *Infrastructuring Publics*, edited by Matthias Korn, Wolfgang Reißmann, Tobias Röhl, and David Sittler, 11–47. Wiesbaden: Springer. https://doi.org/10.1007/978-3-658-20725-0_2.

Laguna, Albert Sergio. 2017. *Diversión: Play and Popular Culture in Cuban America*. New York: New York University Press.

Langley, Paul, and Andrew Leyshon. 2017. "Platform Capitalism: The Intermediation and Capitalization of Digital Economic Circulation." *Finance and Society* 3 (1): 11–31. https://doi.org/10.2218/finsoc.v3i1.1936.

Larkin, Brian. 2004. "Degraded Images, Distorted Sounds: Nigerian Video and the Infrastructure of Piracy." *Public Culture* 16 (2): 289–314. https://doi.org/10.1215/08992363-16-2-289.

———. 2008. *Signal and Noise: Media, Infrastructure, and Urban Culture in Nigeria*. Durham, NC: Duke University Press. https://doi.org/10.1215/9780822389316.

———. 2013. "The Politics and Poetics of Infrastructure." *Annual Review of Anthropology* 42 (1): 327–43. https://doi.org/10.1146/annurev-anthro-092412-155522.

Lea, Tess, and Elizabeth A. Povinelli. 2018. "Karrabing: An Essay in Keywords." *Visual Anthropology Review* 34 (1): 36–46. https://doi.org/10.1111/var.12151.

Ledeneva, Alena V. 1998. *Russia's Economy of Favours: Blat, Networking and Informal Exchange*. Cambridge: Cambridge University Press.

Lee, Ashley. 2018. "Invisible Networked Publics and Hidden Contention: Youth Activism and Social Media Tactics under Repression." *New Media & Society* 20 (11): 4095–4115. https://doi.org/10.1177/1461444818768063.

———. 2022. "Hybrid Activism under the Radar: Surveillance and Resistance among Marginalized Youth Activists in the United States and Canada." *New Media & Society* 26 (7): 3833–53. https://doi.org/10.1177/14614448221105847.

LeoGrande, William M. 2023. "How the Cuban Military Became an Economic Powerhouse." In *Contemporary Cuba: The Post-Castro Era*, edited by Hope Bastian, Philip Brenner, John M. Kirk, and William M. LeoGrande, 176–84. Lanham, MD: Rowman & Littlefield.

Levine, Mike. 2021. "Sounding El Paquete: The Local and Transnational Routes of an Afro-Cuban Repartero." *Cuban Studies* 50: 139–60.

Lodi-Ribeiro, Gerson, ed. 2012. *Solarpunk: Histórias ecológicas e fantásticas em um mundo sustentável*. São Paulo: EditoraDraco.

López, Juan J. 1999. "Implications of the U.S. Economic Embargo for a Political Transition in Cuba." *Cuban Studies* 28: 40–69.

López Moya, Natalia. 2022. "Pese a Su Monopolio, La Cubana Etecsa Se Queda Sin Recursos." *Havana Times*, October 22, 2022. https://havanatimesenespanol.org/reportajes/pese-a-su-monopolio-la-cubana-etecsa-se-queda-sin-recursos/.

Luna Cárdenas, Offray Vladimir. 2019. "Codiseñar Autonomías: Artefactos Digitales Amoldables, Hacktivismo y Ciudadanías." PhD thesis, Universidad de Caldas.

Mandel, Ruth, and Caroline Humphrey, eds. 2002. *Markets and Moralities: Ethnographies of Postsocialism*. Oxford: Routledge.

Mann, Michael. 1984. "The Autonomous Power of the State: Its Origins, Mechanisms and Results." *European Journal of Sociology / Archives Européennes de Sociologie / Europäisches Archiv Für Soziologie* 25 (2): 185–213.

Marche, Guillaume. 2012. "Why Infrapolitics Matters." *Revue française d'études américaines* 131 (1): 3–18. https://doi.org/10.3917/rfea.131.0003.

Marcus, George. 2010. "Para-Sites: A Proto-Prototyping Culture of Method?" *Limn* 1 (0). https://escholarship.org/uc/item/4wr77190.

Mesa Cumbrera, Osmara, Lázara Yolanda Carrazana Fuentes, Dialvys Rodríguez Hernández, Martin Holbraad, Isabel Reyes Mora, and María Regina Cano Orúe. 2020. "State and Life in Cuba: Calibrating Ideals and Realities in a State-Socialist System for Food Provision." *Social Anthropology* 28 (4): 803–26. https://doi.org/10.1111/1469-8676.12961.

Mesa-Lago, Carmelo, and Jorge Pérez-López. 2005. *Cuba's Aborted Reform: Socioeconomic Effects, International Comparisons, and Transition Policies*. Gainesville: University Press of Florida.

———. 2013. *Cuba under Raul Castro: Assessing the Reforms*. Boulder, CO: Lynne Rienner.

Ministerio de Communicaciones, República de Cuba. 2015. "Estrategia Nacional para el desarollo de la infraestructura de conectividad de Banda Ancha en Cuba." https://www.academia.edu/14547523/ESTRATEGIA_NACIONAL_PARA_EL_DESARROLLO_DE_LA_INFRAESTRUCTURA_DE_CONECTIVIDAD_DE_BANDA_ANCHA_EN_CUBA.

Ministerio de Justicia, República de Cuba. 2008. *Gaceta Oficial de la República de Cuba* 60 (November 13, 2008).

Mitchell, Timothy. 1990. "Everyday Metaphors of Power." *Theory and Society* 19 (5): 545–77. https://doi.org/10.1007/bf00147026.

Monk, Simon. 2016. *The Maker's Guide to the Zombie Apocalypse: Defend Your Base with Simple Circuits, Arduino, and Raspberry Pi*. San Francisco: No Starch Press.

Moreiras, Alberto. 2021. *Infrapolitics: A Handbook*. New York: Fordham University Press.

Morley, Morris H., and Chris McGillion. 2002. *Unfinished Business: America and Cuba after the Cold War, 1989–2001*. Cambridge: Cambridge University Press.

Morozov, Evgeny. 2011. *The Net Delusion: How Not to Liberate the World*. New York: Penguin.

Morris, Jeremy, Andrei Semenov, and Regina Smyth. 2023. *Varieties of Russian Activism: State-Society Contestation in Everyday Life*. Bloomington: Indiana University Press.

Mota, Erick J. 2010. *Habana Underguater, La Novela*. Scotts Valley, CA: CreateSpace.

Mouffe, Chantal. 2005. *On the Political*. New York: Routledge.

Natvig, Anne. 2021. "Perceptions of and Strategies for Autonomy among Journalists Working for Cuban State Media." In *Cuba's Digital Revolution: Citizen Innovation and State Policy*, edited by Ted A. Henken and Sara García Santamaria, 200–218. Miami: University Press of Florida. https://doi.org/10.2307/j.ctvlmvw94t.

Nayyar, Rajat, and Magdalena Kazubowski-Houston. 2020. "Talking Uncertainty: An Elephant in the Room." *Emergent Futures CoLab* (blog), July 15, 2020. https://www.urgentemergent.org/talking-uncertainty/kazubowski-houston.

Nepstad, Sharon Erickson. 2011. *Nonviolent Revolutions: Civil Resistance in the Late 20th Century*. Oxford: Oxford University Press.

Nijssen, Edwin J., and Susan P. Douglas. 2011. "Consumer World-Mindedness and Attitudes toward Product Positioning in Advertising: An Examination of Global versus Foreign versus Local Positioning." *Journal of International Marketing* 19 (3): 113–33. https://doi.org/10.1509/jimk.19.3.113.

Oficina Nacional de Estatistica e Informacíon (ONEI). 2015. "Salario Medio En Cifras." http://www.one.cu/salariomedioencifras 2015.htm.

———. 2020. "Anuario Estadástico de Cuba 2019. Empleo y Salarios." http://www.onei.gob.cu/sites/default/files/07_empleo_y_salario_2019_sitio.pdf.

Oramas Pérez, Silvia. 2016. "BUSCANDO SENAL—Acerca de Los Usos Sociales de La Red WIFI_ETECSA Que Realizan Los Usuarios Que Acceden Desde El Parque de 51 En El Municipio Habanero de La Lisa." Thesis, Universidad de la Habana.

Padrón Hernández, Maria. 2012. "Beans and Roses: Everyday Economies and Morality in Contemporary Havana, Cuba." PhD thesis, University of Gothenburg.

Pañellas Álvarez, Daybel. 2021. "Culturas Juveniles: Los TEAMS." In *Identidades Juveniles En Cuba. Claves Para Un Diálogo*, edited by Yoannia Pulgarón Garzón and Ana Isabel Peñate Leiva, 149–68. Havana: Publicaciones Acuario.

Parks, Lisa, and Nicole Starosielski, eds. 2015. *Signal Traffic: Critical Studies of Media Infrastructures*. Champaign: University of Illinois Press.

Patico, Jennifer. 2008. *Consumption and Social Change in a Post-Soviet Middle Class*. Stanford, CA: Stanford University Press.

Pedrito el Paketero, dir. 2019a. "¿Cuál Es El PoRNo Más RARO Que Has Visto? ☞ 🖉 | 10CR." https://www.youtube.com/watch?v=CknFCjNWNhE.

———, dir. 2019b. "¿Están DE ACUERDO Los Cubanos Con Su SISTEMA? | ¿SON FELICES? | 10CR." https://www.youtube.com/watch?v=bLyy-VpAG0Q.

Perry, Marc D. 2015. *Negro Soy Yo: Hip Hop and Raced Citizenship in Neoliberal Cuba*. Durham, NC: Duke University Press.

Pertierra, Anna Cristina. 2009. "Private Pleasures: Watching Videos in Post-Soviet Cuba." *International Journal of Cultural Studies* 12 (2): 113–30.

———. 2011. *Cuba: The Struggle for Consumption*. Coconut Creek, FL: Caribbean Studies Press.

———. 2012. "If They Show *Prison Break* in the United States on a Wednesday, by Thursday It Is Here: Mobile Media Networks in Twenty-First-Century Cuba." *Television & New Media* 13 (5): 399–414.

Plantin, Jean-Christophe, Carl Lagoze, Paul N. Edwards, and Christian Sandvig. 2018. "Infrastructure Studies Meet Platform Studies in the Age of Google and Facebook." *New Media & Society* 20 (1): 293–310. https://doi.org/10.1177/1461444816661553.

Plattner, Marc F., and Larry Diamond. 2012. *Liberation Technology: Social Media and the Struggle for Democracy*. Baltimore: Johns Hopkins University Press.

Ponte, Antonio José. 2005. *Un Arte de Hacer Ruinas y Otros Cuentos*. Mexico City: Colección Aula Atlántica.

———. 2007. *La Fiesta Vigilada*. Barcelona: Editorial Anagrama.

Portes, Alejandro. 2007. "The Cuban-American Political Machine: Reflections on Its Origins and Perpetuation." In *Debating Cuban Exceptionalism*, edited by Bert Hoffmann and Laurence Whitehead, 123–37. New York: Palgrave Macmillan. https://doi.org/10.1007/978-1-137-12353-4_7.

Postigo, Hector. 2007. "Of Mods and Modders: Chasing Down the Value of Fan-Based Digital Game Modifications." *Games and Culture* 2 (4): 300–313. https://doi.org/10.1177/1555412007307955.

Postill, John. 2008. "Localizing the Internet beyond Communities and Networks." *New Media & Society* 10 (3): 413–31. https://doi.org/10.1177/1461444808089416.

Powell, Kathy. 2008. "Neoliberalism, the Special Period and Solidarity in Cuba." *Critique of Anthropology* 28 (2): 177–97. https://doi.org/10.1177/0308275X08090545.

Press, Larry. 1996. "Cuban Telecommunication Infrastructure and Investment." In *Proceedings of the Sixth Annual Meeting of the Association for the Study of the Cuban Economy: Cuba in Transition*, vol. 6, 145–54. Association for the Study of the Cuban Economy.

———. 2011. "The Past, Present, and Future of the Internet in Cuba." In *Proceedings of the Meeting of the Association for the Study of the Cuban Economy*, vol. 21, 186–93. Association for the Study of the Cuban Economy.

Pujol, Eduardo E., Will Scott, Eric Wustrow, and J. Alex Halderman. 2017. "Initial Measurements of the Cuban Street Network." In *Proceedings of the 2017 Internet Measurement Conference*, 318–24. New York: Association for Computing Machinery. https://doi.org/10.1145/3131365.3131395.

Pulgarón Garzón, Yoannia. 2021. "Identificaciones y Pertenencias. Revisitando Las Culturas Juveniles En Cuba." In *Identidades Juveniles En Cuba. Claves Para Un Diálogo*, edited by Yoannia Pulgarón Garzón and Ana Isabel Peñate Leiva, 113–48. Havana: Publicaciones Acuario.

Rai, Amit S. 2019. *Jugaad Time: Ecologies of Everyday Hacking in India*. Durham, NC: Duke University Press.

Rancière, Jacques. 2010. *Dissensus: On Politics and Aesthetics*. Translated by Steve Corcoran. London: Continuum.

Recio Silva, Milena. 2013. "La Hora de Los Desconectados. Evaluación Del Diseño de La Política de 'Acceso Social' a Internet En Cuba En Un Contexto de Cambios." Clacso Respositorio Digital. https://biblioteca-repositorio.clacso.edu.ar/handle/CLACSO/10803.

Redondo, Juan C. Toledano. 2005. "From Socialist Realism to Anarchist Capitalism: Cuban Cyberpunk." *Science Fiction Studies* 32 (3): 442–66.

———. 2019. "Cuba's Cyberpunk Histories." In *The Routledge Companion to Cyberpunk Culture*, edited by Lars Schmeink, Anna McFarlane, and Graham J. Murphy, 395–400. London: Routledge.

Ritter, Archibald R. M., and Ted A. Henken. 2015. *Entrepreneurial Cuba: The Changing Policy Landscape*. Boulder, CO: FirstForumPress.

Rodgers, Dennis, and Bruce O'Neill. 2012. "Introduction: Infrastructural Violence: Introduction to the Special Issue." *Ethnography* 13 (4): 401–12.

Rodríguez Fernández, Fidel A. 2019. "Conexiones comunes: Sobre los usos de las redes autónomas de videojuegos en La Habana y el caso SNET." *IC: Revista Científica de Información y Comunicación* 16: 391–414.

Rogers, Richard. 2019. *Doing Digital Methods*. London: Sage.

Romay Guerra, Zuleica. 2021. "Grietas En La Pared: Una Mirada al Contexto Social Del 11J." In *The Road Ahead—Cuba after the July 11 Protests*, edited by William M. LeoGrande, John M. Kirk, and Philip Brenner. Washington, DC: American University. https://www.american.edu/centers/latin-american-latino-studies/cuba-after-the-july-11-protests-romay-guerra.cfm.

Rosendahl, Mona. 1997. *Inside the Revolution: Everyday Life in Socialist Cuba*. Ithaca, NY: Cornell University Press.

Sanchez, Peter M. and Kathleen M. Adams. 2008. "The Janus-Faced Character of Tourism in Cuba." *Annals of Tourism Research* 35(1): 27–46.

Sawyer, Mark Q. 2005. *Racial Politics in Post-Revolutionary Cuba*. Cambridge: Cambridge University Press.

Schwenkel, Christina. 2015. "Spectacular Infrastructure and Its Breakdown in Socialist Vietnam." *American Ethnologist* 42 (3): 520–34. https://doi.org/10.1111/amet.12145.

Scott, James C. 1985. *Weapons of the Weak: Everyday Forms of Peasant Resistance*. New Haven, CT: Yale University Press.

———. 1990. *Domination and the Arts of Resistance: Hidden Transcripts*. New Haven, CT: Yale University Press.

———. 1998. *Seeing Like a State: How Certain Schemes to Improve the Human Condition Have Failed*. New Haven, CT: Yale University Press.

———. 2012. "Infrapolitics and Mobilizations: A Response by James C. Scott." *Revue française d'études américaines* 131 (1): 112–17. https://doi.org/10.3917/rfea.131.0112.

Simone, AbdouMaliq. 2004. "People as Infrastructure: Intersecting Fragments in Johannesburg." *Public Culture* 16 (3): 407–29.

Srnicek, Nick. 2017. *Platform Capitalism*. Malden, MA: Polity.

Star, Susan Leigh. 1991. "The Sociology of the Invisible: The Primacy of Work in the Writings of Anselm Strauss." In *Social Organization and Social Process: Essays in Honor of Anselm Strauss*, edited by David R. Maines, 265–83. Hawthorne, NY: Aldine de Gruyter.

———. 1999. "The Ethnography of Infrastructure." *American Behavioral Scientist* 43 (3): 377–91. https://doi.org/10.1177/00027649921955326.

Star, Susan Leigh, and Geoffrey C. Bowker. 2002. "How to Infrastructure." In *Handbook of New Media: Social Shaping and Consequences of ICTs*, edited by Leah A. Lievrouw and Sonia Livingstone, 151–62. London: Sage. https://doi.org/10.4135/9781848608245.

Star, Susan Leigh, and Karen Ruhleder. 1996. "Steps toward an Ecology of Infrastructure: Design and Access for Large Information Spaces." *Information Systems Research* 7 (1): 111–34. https://doi.org/10.1287/isre.7.1.111.

Stătică, Iulia. 2019. "Socialist Domestic Infrastructures and the Politics of the Body: Bucharest and Havana." In *The Oxford Handbook of Communist Visual Cultures*, edited by Aga Skrodzka, Xiaoning Lu, and Katarzyna Marciniak, 18–43. Oxford: Oxford University Press. https://doi.org/10.1093/oxfordhb/9780190885533.013.37.

Statista. 2023. "Internet and Social Media Users in the World 2023." May 22, 2023. https://www.statista.com/statistics/617136/digital-population-worldwide/.

Storey, Angela D. 2021. "Implicit or Illicit? Self-Made Infrastructure, Household Waters, and the Materiality of Belonging in Cape Town." *Water Alternatives* 14 (1): 79–96.

Strauss, Anselm. 1985. "Work and the Division of Labor." *Sociological Quarterly* 26 (1): 1–19. https://doi.org/10.1111/j.1533-8525.1985.tb00212.x.

Suchman, Lucy. 1996. "Supporting Articulation Work." In *Computerization and Controversy*, 2nd ed., edited by Rob Kling, 407–23. Boston: Morgan Kaufmann. https://doi.org/10.1016/B978-0-12-415040-9.50118-4.

Takaragawa, Stephanie, Trudi Lynn Smith, Kate Hennessy, Patricia Alvarez Astacio, Jenny Chio, Coleman Nye, and Shalini Shankar. 2019. "Bad Habitus: Anthropology in the Age of the Multimodal." *American Anthropologist* 121 (2): 517–24. https://doi.org/10.1111/aman.13265.

Tankha, Mrinalini. 2021. "Detained Settlements: The Infrastructures and Temporalities of Digital Financial Transactions between the United States and Cuba." *Economic Anthropology* 8 (1): 133–47. https://doi.org/10.1002/sea2.12188.

Tedesco, Laura. 2018. "De militares a gerentes las Fuerzas Armadas Revolucionarias en Cuba." *Nueva Sociedad* 278: 111–18.

Thiemann, Louis, and Claudia Mare. 2021. "Multiple Economies and Everyday Resistance in Cuba: A Bottom-Up Transition." In *Social Policies and Institutional Reform in Post-COVID Cuba*, edited by Bert Hoffmann, 183–206. Leverkusen-Opladen: Verlag Barbara Budrich.

Thompson, E. P. 1971. "The Moral Economy of the English Crowd in the Eighteenth Century." *Past & Present* 50: 76–136.

Tufekci, Zeynep. 2017. *Twitter and Tear Gas: The Power and Fragility of Networked Protest*. New Haven, CT: Yale University Press.

USAID. 2021. "Supporting Local Civil Society and Human Rights in Cuba." July 30, 2021. https://www.grants.gov/web/grants/view-opportunity.html?oppId=334516.

U.S. Congress 1966. The Cuban Adjustment Act, Pub. L. No. 89-732, enacted November 2, 1966, Amended 1995. www.govinfo.gov/content/pkg/STATUTE-80/pdf/STATUTE-80-Pg1161.pdf.

———. 1992. Cuban Democracy Act, Pub. L. No. 102-484, enacted October 23, 1992. https://1997-2001.state.gov/www/regions/wha/cuba/democ_act_1992.html.

———. 1996. Cuban Liberty and Democratic Solidarity (LIBERTAD) Act, Pub. L. No. 104-114, enacted December 3, 1996. https://www.congress.gov/bill/104th-congress/house-bill/927#:~:text=Cuban%20Liberty%20and%20Democratic%20Solidarity%20(LIBERTAD)%20Act%20of%201996%20%2D,international%20peace%3B%20(2)%20othe.

Venegas, Cristina. 2010. *Digital Dilemmas: The State, the Individual, and Digital Media in Cuba*. New Brunswick, NJ: Rutgers University Press.

Venkatesan, Soumhya, Laura Bear, Penny Harvey, Sian Lazar, Laura Rival, and AbdouMaliq Simone. 2018. "Attention to Infrastructure Offers a Welcome Reconfiguration of Anthropological Approaches to the Political." *Critique of Anthropology* 38 (1): 3–52. https://doi.org/10.1177/0308275X16683023.

Vicari, Stefania. 2015. "Exploring the Cuban Blogosphere: Discourse Networks and Informal Politics." *New Media & Society* 17 (9): 1492–1512. https://doi.org/10.1177/1461444814529285.

Vickery, Jacqueline Ryan, and Tracy Everbach. 2018. *Mediating Misogyny: Gender, Technology, and Harassment*. New York: Palgrave Macmillan.

Victor, Hestia. 2019. "'There Is Life in This Place': 'DIY Formalisation,' Buoyant Life and Citizenship in Marikana Informal Settlement, Potchefstroom, South Africa." *Anthropology Southern Africa* 42 (4): 302–15. https://doi.org/10.1080/23323256.2019.1639522.

Vidal Alejandro, Pavel, and Omar Everleny Pérez Villanueva. 2022. "Inflation in Cuba and the Economy's Potential Recovery." *Cuba Capacity Building Project. Horizonte Cubano / Cuban Horizon* (blog), October 8, 2022. https://horizontecubano.law.columbia.edu/news/inflation-cuba-and-economys-potential-recovery-part-1.

von Schnitzler, Antina. 2016. *Democracy's Infrastructure: Techno-Politics and Protest after Apartheid*. Princeton, NJ: Princeton University Press.

Weinreb, Amelia Rosenberg. 2009. *Cuba in the Shadow of Change: Daily Life in the Twilight of the Revolution*. Miami: University Press of Florida.

Weiss, Margot. 2021. "The Interlocutor Slot: Citing, Crediting, Cotheorizing, and the Problem of Ethnographic Expertise." *American Anthropologist* 123 (4): 948–53. https://doi.org/10.1111/aman.13639.

Weist, Julia, and Nestor Siré. 2020. "Proyecto DATA." *Triple Canopy*, October 30, 2020. https://canopycanopycanopy.com/contents/proyecto-data?q=data.

Westmoreland, Mark R. 2022. "Multimodality: Reshaping Anthropology." *Annual Review of Anthropology* 51 (1): 173–94. https://doi.org/10.1146/annurev-anthro-121319-071409.

Wilson, Marisa. 2013. *Everyday Moral Economies: Food, Politics and Scale in Cuba*. Chichester: Wiley-Blackwell.

Winthereik, Brit Ross, and Ayo Wahlberg. 2022. "Infrastructures, Linkages, and Livelihoods." In *Handbook for the Anthropology of Technology*, edited by Maja Hojer Bruun, Ayo Wahlberg, Rachel Douglas-Jones, Cathrine Hasse, Klaus Hoeyer, Dorthe Brogård Kristensen, and Brit Ross Winthereik, 673–87. Singapore: Palgrave Macmillan. https://doi.org/10.1007/978-981-16-7084-8_34.

Wylie, Lana, and Lisa Glidden. 2013. "The 'Cuban Spring' Fallacy: The Current Incarnation of a Persistent Narrative." *International Journal of Cuban Studies* 5 (2): 140–67. https://doi.org/10.13169/intejcubastud.5.2.0140.

Yaffe, Helen. 2021. "Day Zero: How and Why Cuba Unified Its Dual Currency System." *LSE Latin America and Caribbean Blog*, February 10, 2021. https://blogs.lse.ac.uk/latamcaribbean/.

Yang, Peidong, Lijun Tang, and Xuan Wang. 2015. "Diaosi as Infrapolitics: Scatological Tropes, Identity-Making and Cultural Intimacy on China's Internet." *Media, Culture & Society* 37 (2): 197–214. https://doi.org/10.1177/0163443714557980.

Zuboff, Shoshana. 2019. *The Age of Surveillance Capitalism: The Fight for a Human Future at the New Frontier of Power*. New York: PublicAffairs.

INDEX

GPSR Authorized Representative: Easy Access System Europe - Mustamäe tee
50, 10621 Tallinn, Estonia, gpsr.requests@easproject.com

www.ingramcontent.com/pod-product-compliance
Lightning Source LLC
Chambersburg PA
CBHW030732280326
41926CB00086B/1182